Mathematical Modelling Techniques

This book is based on lecture materials originally developed by Harald E. Krogstad for the MSc course *Mathematical Modelling* at the Norwegian University of Science and Technology (NTNU) and later adapted for the MASTMO MSc program at Hawassa University, Ethiopia. Following Prof. Krogstad's passing in 2020, the project was continued by Mohammed Yiha Dawed and Patrick M. Tchepmo Djomegni, with contributions in mathematical epidemiology from Julien Arino. The text offers a structured and comprehensive introduction to mathematical modelling, blending classical methods with original content. Its goal is to equip beginning graduate students and advanced undergraduates in applied mathematics, statistics, and related fields with the tools to formulate, analyze, and interpret models across natural sciences, engineering, and biology.

What distinguishes this book is its balance of theory and application, bridging rigorous mathematics with practical modelling problems. Students learn both the conceptual foundations and computational approaches, preparing them to tackle real-world challenges.

Key Features Include:

- Introduction to dimensional analysis, scaling, and perturbation methods.
- Equilibrium, bifurcation, and hysteresis analysis.
- Classical and generalized population and epidemic models, including logistic, Lotka-Volterra, and SIR models.
- Early introduction to partial differential equations through conservation-based modelling.
- Applications to road traffic, mechanics, diffusion, and small project-based exercises.
- Worked examples and chapter-end exercises for self-study.
- Concise "first aid" guide for solving first-order quasi-linear PDEs.

This book is ideal for beginning graduate students and advanced undergraduates in applied mathematics, statistics, and related disciplines. Its combination of theory, applied examples, and project-based exercises makes it a valuable resource for courses as well as for independent study.

Mathematical Modelling Techniques

Harald E. Krogstad, Mohammed Yiha Dawed,
Patrick M. Tchepmo Djomegni, and Julien Arino

CRC Press
Taylor & Francis Group
Boca Raton London New York

CRC Press is an imprint of the
Taylor & Francis Group, an **informa** business

A CHAPMAN & HALL BOOK

First edition published 2026
by CRC Press
2385 NW Executive Center Drive, Suite 320, Boca Raton FL 33431

and by CRC Press
4 Park Square, Milton Park, Abingdon, Oxon, OX14 4RN

CRC Press is an imprint of Taylor & Francis Group, LLC

ISBN: 978-1-041-21009-2 (hbk)
ISBN: 978-1-041-21007-8 (pbk)
ISBN: 978-1-003-72520-6 (ebk)

DOI: 10.1201/9781003725206

Typeset in Alegreya-Regular font
by KnowledgeWorks Global Ltd.

Publisher's note: This book has been prepared from camera-ready copy provided by the authors.

This book is dedicated to the memory of Professor Harald E. Krogstad (1945–2020), a distinguished applied mathematician, engineer, inspiring teacher, and mentor.

His clarity of thought, passion for mathematical modelling, and commitment to education shaped generations of students and laid the foundation for this work.

Contents

Preface

Background and Content

The material presented in this book originates from the long teaching experience of Professor Harald E. Krogstad, who developed it while lecturing on the MSc course Mathematical Modelling at the Norwegian University of Science and Technology (NTNU), Trondheim. Initially written in Norwegian as a collection of notes and handouts, the text was later structured, translated into English, and used in the MASTMO MSc program at Hawassa University, Ethiopia. After Professor Krogstad's passing in 2020, Mohammed Yiha Dawed (Hawassa University, Ethiopia) and Patrick Tchepmo Djomegni (North-West University, South Africa) continued the project, with Julien Arino (University of Manitoba, Canada) contributing new content in mathematical epidemiology.

This book is aimed at beginning graduate students and advanced undergraduates in applied mathematics and statistics. It introduces fundamental modelling tools that are widely applicable in natural sciences, engineering, and biology. Students are expected to be familiar with calculus, linear algebra, basic ordinary and partial differential equations, and numerical software such as Octave (Matlab) or R. While some chapters are adapted from classical modelling texts such as Lin and Segel [30], Logan [31], and Kermack and McKendrick [26], much of the material is original. We aim to bring a distinctive flavoor, especially for teaching mathematical modelling concepts at the advanced undergraduate and graduate levels.

The book begins in Chapter 1 with dimensional analysis, laying the foundation for understanding how physical laws can be reduced to dimensionless forms. Buckingham's Pi-theorem is introduced, followed by diverse applications ranging from fluid flows and water waves to biological systems such as food chain models. Scaling is the subject of Chapter 2, where students learn how to identify dominant effects and simplify equations. Real-world examples, including turbulence, geometric similarity in animals, and athletic activities such as running and weightlifting, illustrate how scaling provides powerful insights.

Chapters 3 and 4 develop the essential tools of perturbation and stability analysis. Regular and singular perturbation methods are introduced with applications to projectile motion, chemical kinetics, and enzyme reactions.

Stability theory is then discussed through linear and nonlinear systems, bifurcation analysis, and phenomena such as hysteresis and limit cycles. These techniques equip students to move beyond exact solutions, providing approximations and qualitative insights into complex systems.

In Chapter 5, attention turns to population models, beginning with single-species dynamics (Malthusian growth, logistic growth, and Allee effects) and extending to multi-species systems, including competition, mutualism, and predator–prey dynamics. Chapter 6 focuses on epidemiological modelling, presenting the Kermack–McKendrick and SIR models, along with concepts such as the basic reproduction number, herd immunity, and vaccination. These models form the foundation of modern mathematical epidemiology.

From Chapter 7 onwards, the book shifts to modelling based on conservation principles, introducing density, flux, and sources/sinks. This framework leads to applications in mechanics, diffusion, and transport phenomena. Chapter 8 develops models of road traffic, from kinematic theory to shock formation and instabilities, while Chapter 9 formalizes conservation laws of mass, momentum, and energy with examples such as flood waves and pipe flows. Chapter 10 extends this to diffusion and convection, with applications in chemical reactors, nuclear accidents, and turbulence.

To strengthen practical understanding, Chapter 11 presents a series of modelling projects ranging from environmental applications (river contamination, lake sedimentation, and water purification) to biomedical and engineering contexts (drug distribution, porous media flow, and traffic control). These projects provide opportunities for students to apply the methods developed in earlier chapters to realistic scenarios. Finally, Chapter 12 offers a concise introduction to first-order quasi-linear PDEs, with a "first aid crash course" for solving them. This serves as a gateway for students preparing to engage with more advanced PDE-based modelling.

Throughout the text, solved examples illustrate the techniques, while exercises at the end of each chapter provide opportunities for practice and deeper engagement. The balance between theory and application means that the material may sometimes feel "too mathematical" for students from the sciences and "too applied" for those with a strong mathematical orientation. Yet this balance reflects the true nature of mathematical modelling, which thrives at the intersection of rigorous mathematics and real-world application.

How to Use This Book

This book is designed to be flexible for both self-study and classroom teaching:

- For students: Begin with Chapters 1 and 2 to build intuition for dimensional analysis and scaling. These chapters are accessible and provide tools you will use throughout the book. Work carefully through the solved examples, and attempt the exercises to reinforce concepts. In later chapters, focus on the case studies and modelling projects to see theory applied to real-world systems.

- For instructors: The modular structure allows chapters to be taught independently or integrated into existing courses. For example, Chapters 1–4 can form a short course on modelling fundamentals, while Chapters 5–6 may be emphasized in a course on mathematical biology. The projects in Chapter 11 are particularly suited for group work, seminars, or as the basis for term papers.

- For researchers: The text can serve as a quick reference for modelling techniques across disciplines. The emphasis on practical examples and applications makes it useful when approaching interdisciplinary problems in biology, engineering, or the physical sciences.

Readers are encouraged to approach the text actively, working through the mathematics with pencil and paper, and complementing the analysis with numerical experiments in Octave, Matlab, or R. The exercises are not merely practice problems but are intended to deepen understanding and stimulate independent thinking. We hope that this book will not only serve as an introduction to mathematical modelling but also inspire further exploration and discovery in applying mathematics to the challenges of science and society.

Mathematical Modelling in Practice

Mathematics, as practiced in applied science, engineering, and industry, differs in many ways from mathematics as a purely academic discipline. First and foremost, practical mathematics in a non-academic setting typically tends to be quite targeted, where one goes in with all the tools one masters in order to solve or get maximum insight about a problem. Problems are generally interdisciplinary and complex, and at least it will almost never be such that the problems are fully formulated and have a smart and simple solution, as we know it from regular courses in mathematics.

In 1995, a group from the Society of Industrial and Applied Mathematics (SIAM) carried out a survey of industrial and applied mathematicians in the private and public sectors. Both mathematicians and their managers were asked how relevant they considered the university training to be in their current job. A main conclusion from the industry managers' side was the importance of skills in formulating, modelling, and solving problems from diverse and changing areas. Moreover, an interest in and knowledge of applications, as well as experience with computation, are important. But also spoken and written communication skills and an ability to function in teamwork were considered very important. On the other hand, the applied mathematicians complained that they never got clean and well-formulated problems as they were used to during their studies. The survey and its conclusions are equally, if not more, relevant today.

If we look at the work of the applied mathematician in a project, it typically will involve four stages which may be denoted as *absorbing, modelling, analyzing,* and *communicating.*

Absorbing: This involves understanding and learning about the problem. Typically, the problem will be communicated by clients or collaborators in a non-mathematical language, vague and incomplete, and often with expressions and phrases that are specific to the field and unfamiliar to the mathematician. In this phase of the project, the mathematician often feels or may even be accused of being both a slow learner and worse, but this should not prevent her/him to stop before receiving the maximum amount of information about the problem. In the mathematical modelling course, we shall learn a few techniques that may be used to confront the problem providers. Are they actually able to make all equations dimensionless, and above all, do they know the scales of the problem? This is actually commented on the SIAM report: Some of the industry leaders said that even they did not always have so much help from the mathematicians, they were very clever to ask the right and important questions!

Today a significant part of this phase would be to find relevant and reliable information on the internet.

Modelling: For the modelling, one must understand what is going on and translate this into a mathematical form. Dimensional analysis, scaling, and conservation principles are general techniques that help us with this. The modelling is often iterative, with a lot of trial and error.

Analysing: Qualitative analysis techniques of models used such as stability, bifurcation, and perturbation analysis, is important. Quantitative analysis helped to find analytic solutions. Today has one also the opportunity to do experiments by means of a computer. Of course, one cannot always generalize from such simple experiments. However, this will sometimes initiate a need for more extensive numerical solutions (often denoted Computational Science and Engineering, *CSE*) for most of the problems one encounters in practice. But it is absolutely necessary to have a thorough modelling and analysis *before* starting the extensive work which is to arrange and carry out large-scale numerical computations.

Communicating: After the modelling and the results of the analysis, it is time for a presentation to the clients. These are generally not mathematicians, and the mathematician must realize that the traditional academic presentation used in a scientific thesis or publication may be totally useless in this context. It is something else if the problem is mathematically interesting and one is writing a scientific report aimed at colleagues. However, when this is said, it can also be added that one may well keep the precision and logic applied in mathematics, even if the abstraction level is reduced.

Good luck!

Acknowledgments

We thank the students of the Norwegian University of Science and Technology (Norway) and Hawassa University (Ethiopia) whose engagement with the original course materials shaped the evolution of this book. Their curiosity, feedback, and challenges provided invaluable insights into how mathematical modelling can best be taught and learned.

Patrick M. Tchepmo Djomegni gratefully acknowledges the financial support provided by the National Institute for Theoretical and Computational Sciences (NITHeCS), the Association pour la Promotion Scientifique de l'Afrique (APSA), and the research entity Pure and Applied Analytics (PAA) at North-West University. This support made it possible to undertake a six-week research visit to Mohamed Yiha Dawed in Ethiopia, during which substantial progress was achieved towards the completion of this book. I extend my heartfelt gratitude to my wife, Arlette Nwanmou Youkoumbou, and to my children, Joshua, Daniella, Gloria, and Ange, for their unwavering understanding, love, and support.

Authors

Harald E. Krogstad was a distinguished Norwegian applied mathematician and Professor Emeritus at the Norwegian University of Science and Technology, Trondheim, where he earned his PhD in 1969. He made seminal contributions to signal processing, ocean wave dynamics, and numerical simulations. Author of over 110 publications, his expertise spanned wave spectra, digital signal processing, and stochastic analysis.

Mohammed Yiha Dawed is an Associate Professor of Applied Mathematics at Hawassa University, Ethiopia, where he earned his PhD in 2017. He previously served as Head of the Department of Mathematics and contributed to Ethiopia's national curriculum by editing Grade 9 and 10 textbooks and teachers' guides. His research focuses on differential equations applied to epidemiology, ecology, and eco-epidemiology.

Patrick M. Tchepmo Djomegni is an Associate Professor at North-West University and a C2-rated researcher by South Africa's NRF. He earned his PhD in Applied Mathematics from the University of KwaZulu-Natal in 2015. Founder of 3MC, he serves on CIMPA's Governing Board and co-leads the CNRS-AFRICA JPR FANE-MATH-PE chair (2024-2027). His research centers on differential equations in epidemiology, ecology, and oncology.

Julien Arino is a Professor with the Department of Mathematics at the University of Manitoba, in Winnipeg, Canada. He obtained his PhD at INRIA Sophia Antipolis, France, in 2001. He is a mathematical biologist working mostly in epidemiology and ecology. He is particularly interested in the relationship between population movement and the spatio-temporal spread of infectious pathogens.

1

Dimensional Analysis

1.1 Basis of Dimensional Analysis

Dimensional Analysis is a technique based on two simple axioms about nature:

- *All relations between physical quantities must be dimensionally correct.*

- *No physical relation should depend on any particular set of units.*

Even if these axioms sound trivial and obvious, they lead to a powerful, simple, and quite useful tool in mathematical modelling. At the same time, it illustrates how important it is to uncover the mathematical essence in two general and apparently vague statements. Let us therefore now investigate the axioms in somewhat more detail. A physical quantity has

- *dimension,*

- *unit,*

- *numerical value.*

Dimensions can be *Length, Time, Mass,* or combinations of them. The dimension of a physical quantity is given once and for all and will never change. When we work with physical quantities, we need units, and as we know, there is a huge amount of units. For length we have cm, m, km, foot, inch, etc. When the unit changes, the numerical value of the quantity also changes, as illustrated in Figure 1.1.

Let R be a physical quantity. The reader should have observed by now that we have not really defined what a physical quantity is. *Wikipedia* defines a physical quantity as a physical property that can be quantified in terms of numbers. Thus, the mass of Earth is a physical quantity, whereas a dice showing six dots for a rock concert is not. It is convenient to have a notation for the unit of R, and we shall write this as $[R]$. The numerical value of R when we use a certain unit is denoted $v(R)$. This is not a standard notation, but it is convenient for the moment. Hence, R can be written in the form

$$R = v(R)[R].$$

DOI: 10.1201/9781003725206-1

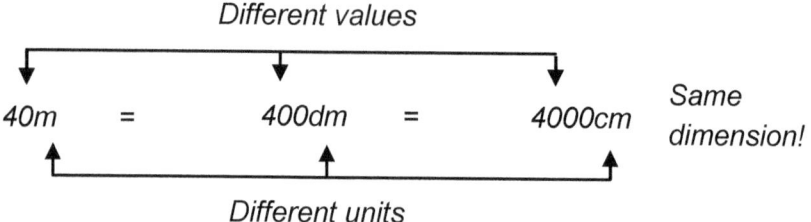

FIGURE 1.1
The numerical value changes when the unit changes, but the physical dimension remains the same.

A physical relation or equation is *dimensionally correct* if each side of an equality has the same dimensions. It is obvious that the well-known formula $S = \frac{1}{2}gt^2$ is dimensionally correct since

$$[S] = \text{m},$$
$$[g] = \text{m/s}^2,$$
$$[t] = \text{s}.$$

On the contrary, you often find, even in textbooks, equations of the form

$$S = 4.9 \times t^2.$$

Such relations should not be used. If you cannot state your equations in a dimensionally correct form, you probably do not have a clear idea of what is going on. In physics there is a set of dimensions forming, so to speak, the *atoms*. None of these depend on the others and the dimensions along with corresponding units in the *International System of Units* (abbreviated the *SI-system*) are listed in Table 1.1. All physical quantities have units which are power combinations of the basic SI-system units (this should actually be considered as the definition of a physical quantity).

1.2   Buckingham's Pi-theorem

Buckingham's Pi-theorem extracts the mathematical content of the two axioms listed above. The somewhat strange name comes from the dimensionless variables that we end up with when we apply the theorem. These are often referred to as π_1, π_2, $\cdots$ (Π is also used instead of π).

Let us look at what *a relationship between physical quantities* means. A relationship is a relation or a formula, that is, an equation that we can write

TABLE 1.1
Fundamental physical units in the SI-system. Exact definitions may be found in Wikipedia.

Quantity	Dimension symbol	SI unit
Mass	M	kilo (kg)
Length	L	meter (m)
Time	T	second (s)
Electric current	I	ampere (A)
Absolute temperature	Θ	kelvin (K)
Amount of substance	N	mole (mol)
Luminous intensity	J	candela (Cd)

in the form
$$\Phi(R_1, R_2, \cdots, R_M) = 0,$$

where Φ is a certain function. We could also write this relation in other ways, e.g.,
$$R_1 = \Psi(R_2, \cdots, R_M).$$

If we choose a set of fundamental units and use these in a consistent way for all the involved variables, we should now have
$$\Phi(v(R_1), v(R_2), \cdots, v(R_M)) = 0.$$

In practice we could, for example, specify $v(R_2), \cdots, v(R_M)$, and then calculate $v(R_1)$ from the relationship. However, it may well happen that there is no valid relationship between $v(R1), \cdots, v(R_M)$. If we consider S and t where $[S] = $ m and $[t] = $ s, there is no valid physical relation that contains *only* these two variables. We need at least one more quantity, as in the relations $S = gt^2/2$ or $S = Vt$. By simply looking at the quantities and their units, it is possible to decide whether they can at all be combined into a sensible relation.

To investigate this further, it is smart to create a so-called *dimension matrix* containing the exponents of the fundamental units in the units for the quantities we have. If $[S] = $ m, $[t] = $ s, and $[g] = $ ms^{-2}, the dimension matrix will be

	S	t	g
m	1	0	1
s	0	1	-2

Let us point out here that we list the familiar units in the first column. Actually, it would be more correct to use universal dimension assignments such as L for *length*, T for *time*, M for *mass*, etc. This is found in many textbooks. The dimension matrix surveys what the dimensions of involved variables are.

Let now, in general, $F_1, F_2, \cdots, F_N$ denote the fundamental units in Table 1. The units of any physical quantity may be expressed by means of these, for example, the unit for *energy* is "kgm^2/s^2". Generally, we may thus write

$$[R_1] = F_1^{a_{11}} F_2^{a_{21}} \cdots F_N^{a_{N1}},$$

$$\vdots$$

$$[R_M] = F_1^{a_{1M}} F_2^{a_{2M}} \cdots F_N^{a_{NM}}.$$

This gives the dimension matrix $\mathbf{A}$:

	R_1	R_2	$\cdots$	R_M
F_1	a_{11}		$\cdots$	a_{1M}
F_2	$\vdots$	$\mathbf{A}$		$\vdots$
$\vdots$			$\ddots$	
F_N	a_{N1}		$\cdots$	a_{NM}

We say that $R_1, \cdots, R_r$ have *independent dimension* if it is impossible to make a (non-trivial) dimensionless combination of $R_1, \cdots, R_r$ of the form

$$R_1^{\lambda_1} R_2^{\lambda_2} \times \cdots \times R_r^{\lambda_r}.$$

In order to see what this means, we determine the unit of this expression:

$$\left[R_1^{\lambda_1} R_2^{\lambda_2} \times \cdots \times R_r^{\lambda_r} \right] = F_1^{a_{11}\lambda_1 + a_{12}\lambda_2 + \cdots a_{1r}\lambda_r} \times$$

$$\times F_2^{a_{21}\lambda_1 + a_{22}\lambda_2 + \cdots a_{2r}\lambda_r} \times \cdots$$

$$\cdots \times F_N^{a_{N1}\lambda_1 + a_{N2}\lambda_2 + \cdots a_{Nr}\lambda_r}.$$

Then, $R_1, \cdots, R_r$ have independent dimensions if the system of equations

$$a_{11}\lambda_1 + a_{12}\lambda_2 + \cdots a_{1r}\lambda_1 = 0,$$
$$a_{21}\lambda_1 + a_{22}\lambda_2 + \cdots a_{2r}\lambda_r = 0,$$

$$\vdots$$

$$a_{N1}\lambda_1 + a_{N2}\lambda_2 + \cdots a_{Nr}\lambda_r = 0,$$

only has the trivial solution

$$\lambda_1 = \lambda_2 = \cdots = \lambda_r = 0.$$

We may recall from the theory of linear equations that this happens if and only if the matrix columns

$$\begin{bmatrix} a_{11} \\ \vdots \\ a_{N1} \end{bmatrix}, \begin{bmatrix} a_{12} \\ \vdots \\ a_{N2} \end{bmatrix}, \cdots, \begin{bmatrix} a_{1r} \\ \vdots \\ a_{Nr} \end{bmatrix}$$

are *linearly independent*. Thus, we have proved that $R_1, \cdots, R_r$ have independent dimensions if and only if the corresponding dimensional matrix has linearly independent columns. Since the column vectors have length N, we must have $r \leq N$ for this to be possible. Going back to the example above, we see that S and t have independent dimensions. The same applies to $\{S, g\}$ and $\{t, g\}$. On the contrary, S, t, and g do not have independent dimensions.

If we now have a general dimension matrix, the maximum number of variables with independent dimensions is equal to the maximum number of columns that are linearly independent. This is known in linear algebra as the *rank* of the matrix. Let us assume that we have organized ourselves so that $R_1, \cdots, R_r$ have independent dimensions, and that r is the rank of A, $r = \text{rank}(A)$. We may assume that $r < M$ (if $r = M$, all quantities have independent dimensions and there will be no non-trivial physical relationship between them). From this assumption, we may use $R_1, \cdots, R_r$ to form combinations involving $R_{r+1}, \cdots, R_M$ such that

$$\pi_1 = R_{r+1}/(R_1^\bullet \times \cdots \times R_r^\bullet),$$
$$\pi_2 = R_{r+2}/(R_1^\bullet \times \cdots \times R_r^\bullet),$$
$$\vdots$$
$$\pi_{M-r} = R_M/(R_1^\bullet \times \cdots \times R_r^\bullet),$$

are dimensionless. Here "$\bullet$" means suitable exponents so as to make the π-s dimensionless. If we then have a relation

$$\Phi(R_1, R_2, \cdots, R_M) = 0,$$

it is possible to replace $R_{r+1}, \cdots, R_M$, and arrange the expression such that we end with a new, but *equivalent* relation,

$$\Psi(R_1, R_2, \cdots, R_r, \pi_1, \cdots, \pi_{M-r}) = 0.$$

We now claim: *If $R_1, \cdots, R_r$ have independent dimensions, it is possible to choose a set of fundamental units such that the numerical values $v(R_1), v(R_2), \cdots, v(R_r)$ become arbitrarily specified positive numbers!*

This is easy to see if we have just one quantity, e.g., $R_1 = 40$m. If we measure R_1 in centimeters, $v(R_1) = 4000$, while measured in kilometers, $v(R_1) = 0.04$, and so on.

If the claim is correct, we have obtained the following interesting situation: Whatever units we decide to use,

$$\Psi(v(R_1), v(R_2), \cdots, v(R_r), v(\pi_1), \cdots, v(\pi_{M-r})) = 0.$$

While the first r variables may take any positive value depending on how we choose the units, the latest $M - r$ variables remain constant during this change. As a function of M variables, Ψ will therefore always be completely unaffected by the values of the r first arguments. In other words, Ψ can really only depend on $\pi_1, \cdots, \pi_{M-r}$! We have thus reduced our relation with M variables, $\Phi(R_1, R_2, \cdots, R_M) = 0$ to a new relation $\Psi(\pi_1, \cdots, \pi_{M-r}) = 0$ with only $M - r$ variables, all of which are dimensionless. Apart from the claim above, which is proved below, we have now obtained

Buckingham's Pi-theorem:
If there exists a (physically proper) relation

$$\Phi(R_1, R_2, \cdots, R_M) = 0$$

between the quantities $R_1, R_2, \cdots, R_M$, there also exists an equivalent relation

$$\Psi(\pi_1, \cdots, \pi_{M-r}) = 0,$$

where r is the rank of the dimension matrix.

Note that the theorem *assumes* states that there is a relationship between $R_1, R_2, \cdots$, and R_M. This has to be ensured, or at least assumed, before we apply the theorem. In fact, Buckingham's Pi-theorem may also prove that no such relation exists. Buckingham's Pi-theorem reduces the number of parameters, and if none of the dimensionless π-s are completely redundant, $M - r$ is also the least possible number of variables we have in our problem.

In the proof above we applied $R_1, R_2, \cdots, R_r$ to create the dimensionless combinations. These variables are often called *core variables*. Usually, there are several possibilities for the core variables, and what is appropriate depends on the problem.

It is easy to set up a formal procedure to determine the π-s (see Sec. 1.3.2 below). This is also found in the textbooks, e.g., [35], but most of the time it is just as easy to find the combinations by a simple inspection. Note that the number of dimensionless variables will be the same regardless of the choice of core variables and combinations.

In the remainder of this section, we shall, for those particularly interested, show the above claim.

We assume therefore that $R_1, \cdots, R_r$ have independent dimensions, and that the dimension matrix

$$\begin{bmatrix} a_{11} & \cdots & a_{1r} \\ \vdots & \ddots & \vdots \\ a_{N1} & \cdots & a_{Nr} \end{bmatrix}$$

has rank r. A well-known proposition from linear algebra says that the *rank of a matrix $\mathbf{A}$ is the same as the rank of the transposed matrix $\mathbf{A}^T$*. It turns out that this sentence is exactly what we need to complete the proof.

Let us assume that, using the fundamental units $F_1, F_2, \cdots, F_N$, R_i has the values $v_F(R_i)$, $i = 1, \cdots, r$. For another set of units for the same dimensions,

$G_1, G_2, \cdots, G_N$, the numerical values will be $v_G(R_i)$, $i = 1, \cdots, r$. Let $x_i = F_i/G_i$. Then,

$$
\begin{aligned}
R_i &= v_F(R_i) F_1^{a_{1i}} F_2^{a_{2i}} \cdots F_N^{a_{Ni}} \\
&= v_F(R_i) x_1^{a_{1i}} G_1^{a_{1i}} x_2^{a_{2i}} G_2^{a_{2i}} \cdots x_N^{a_{Ni}} G_N^{a_{Ni}} \\
&= v_F(R_i) x_1^{a_{1i}} x_2^{a_{2i}} \cdots x_N^{a_{Ni}} G_1^{a_{1i}} G_2^{a_{2i}} \cdots G_N^{a_{Ni}} \\
&= v_G(R_i) G_1^{a_{1i}} G_2^{a_{2i}} \cdots G_N^{a_{Ni}}.
\end{aligned}
$$

Thus,

$$
v_F(R_i) x_1^{a_{1i}} x_2^{a_{2i}} \cdots x_N^{a_{Ni}} = v_G(R_i), \quad i = 1, \cdots, r.
$$

If we take the logarithm of both sides, we end up with a linear system of equations of the form

$$
\begin{aligned}
a_{11} \log(x_1) + \cdots + a_{N1} \log(x_N) &= \log(v_G(R_1)) - \log(v_F(R_1)), \\
a_{12} \log(x_1) + \cdots + a_{N2} \log(x_N) &= \log(v_G(R_2)) - \log(v_F(R_2)),
\end{aligned}
$$

$$
\vdots
$$

$$
a_{1r} \log(x_1) + \cdots + a_{Nr} \log(x_N) = \log(v_G(R_r)) - \log(v_F(R_r)).
$$

We recognize the coefficient matrix in this linear equation system as the *transposed* of the dimension matrix. Since we have r equations and, according to the proposition from linear algebra, r linearly independent columns in the coefficient matrix, this system will have solutions $\log(x_1), \cdots, \log(x_N)$ (not necessarily unique) *regardless* of the choice of the right-hand side. But this means that we may first specify $v_G(R_1), \cdots, v_G(R_r)$ to whatever we want, then find $\log(x_1), \cdots, \log(x_N)$, then $x_1, \cdots, x_N$, and finally select custom G-units, $G_i = F_i/x_i$, $i = 1, \cdots, r$.

1.3   Applications

1.3.1   First atomic bomb explosion

The following example of the use of dimensional analysis has become a classic. The English physicist G. I. Taylor watched an amateur film of the first American atomic bomb explosion in the Nevada desert and measured the radius (r) of the fireball as a function of time (t); see Figure 1.2. He argued that r, apart from t, should depend on the energy (e) that is released in the explosion and the density (ρ) of the air, since the flame-front needs to accelerate the mass of the surrounding air. Thus, he assumed that there is a relation $\Phi(r, t, \rho, e) = 0$, leading to the following dimension matrix

	r	t	ρ	e
kg	0	0	1	1
m	1	0	-3	2
s	0	1	0	-2

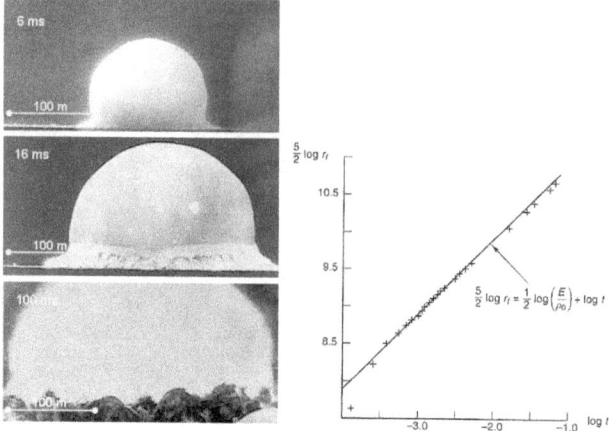

FIGURE 1.2
This series of images from the first atomic bomb explosion shows a fireball growing with time. To the right is shown a copy of the figure G.I. Taylor published, based on the whole sequence of images (see the book by Barenblatt [6] pp. 47–50).

Note that we can find the unit of energy, expressed in terms of the fundamental units, by consulting well-known formulas from physics. For example, we know that energy is force×distance, and that force is mass×acceleration. This gives us

$$[e] = \text{Nm} = (\text{kgm/s}^2)\text{m} = \text{kgm}^2/\text{s}^2.$$

The dimension matrix above has rank 3, and, according to Buckingham, there is $4 - 3 = 1$ dimensionless parameter. Since the equation is simply $\Psi(\pi_1) = 0$, we assume it has a unique solution such that we may write

$$\pi_1 = C,$$

where C is an unknown constant. We may find π_1 by *trial and error*. First,

$$\left[\frac{e}{\rho}\right] = \frac{\text{Nm}}{\text{kg/m}^3} = \frac{\text{kgm}^2}{\text{s}^2}\frac{\text{m}^3}{\text{kg}} = \frac{\text{m}^5}{\text{s}^2},$$

and then we observe that the combination is dimensionless

$$\pi_1 = \frac{e\,t^2}{\rho\,r^5}.$$

This gives us the following simple formula:

$$e = C\rho\frac{r^5}{t^2}.$$

We are not able to determine the constant C, but from the film G.I. Taylor was able to find the ratio r^5/t^2, and by assuming $C = 1$ (best guess!), he got that the released energy $e \approx 10^{14}$ J. It turned out that this was within a factor of 2 of the correct value. The publication of the energy, which was of course "top secret", caused great confusion among the Americans when this was published as a letter in The Times.

1.3.2 Recipe for finding dimensionless combinations

If it is difficult to see the dimensionless combinations directly, it is possible to put up a system of equations for the exponents. The method may be illustrated using the example above. Here we are looking for a dimensionless combination of the form $\pi = e^x \rho^y r^z t^u$ and must therefore determine $\{x, y, z, u\}$. Since we already know the units of the variables,

$$[\pi] = [e^x \rho^y r^z t^u]$$
$$= \left(kg^x m^{2x} s^{-2x}\right) \left(kg^y m^{-3y}\right) (m^z) (s^u)$$
$$= kg^{x+y} m^{2x-3y+z} s^{-2x+u}.$$

Here π should be dimensionless, and therefore,

$$x + y = 0,$$
$$2x - 3y + z = 0,$$
$$-2x + u = 0.$$

There is no unique solution, but if we choose $x = 1$, it follows easily that $y = -1$, $z = -5$, and $u = 2$, in other words, exactly what we already obtained above. There is no reason to use this cumbersome method if it is possible to see the result directly.

1.3.3 Pythagorean theorem

Since the area of a right-angled triangle is completely determined by the length of the hypotenuse (c) and the smallest angle (α_{min}), there must be a relationship between the surface area (A) of the triangle, the length of the hypotenuse, and the angle, of the form

$$\Phi(A, c, \alpha_{min}) = 0,$$

where Φ is a function. It is easy to set up the dimension matrix:

	A	c	α_{min}
m	2	1	0

(An angle is measured in radians, which is a ratio between two lengths and thus dimensionless.) Since the rank of the matrix is 1, there are two dimensionless

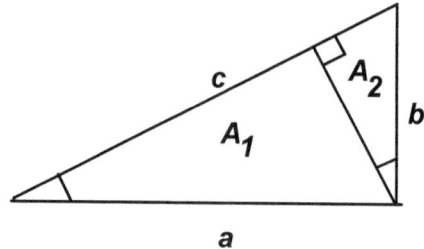

FIGURE 1.3
The area of the big triangle is the sum of the equally shaped smaller triangles.

parameters, A/c^2 and $\alpha_{\min}$, and we end up with a relation $A = c^2 f(\alpha_{\min})$. From Figure 1.3, it is obvious that $A = A_1 + A_2$ for the areas, and therefore $c^2 f(\alpha_{\min}) = a^2 f(\alpha_{\min}) + b^2 f(\alpha_{\min})$, or $c^2 = a^2 + b^2$, which is yet another proof of this famous result. We note that A_1 and A_2 are areas of right-angled triangles and have the same smallest angle ($\alpha_{\min}$). Hence, $A_1 = a^2 f(\alpha_{\min})$ and $A_2 = b^2 f(\alpha_{\min})$.

1.3.4 Fluid flows in tubes

This example, giving an expression for computing the friction experienced by viscous fluids flowing in tubes, is quite famous and very useful in engineering. It requires some background in fluid mechanics.

We shall find an expression for the pressure drop in a fluid flowing through a cylindrical tube which contains a flowing fluid, and we assume that the variables in Table 1.2 are important.

Concerning *viscosity*, μ, this is a proportionality constant between *shear stress (force/area unit)*, for example, σ_{yx}, and changes in the speed per unit length, $\partial u/\partial y$, normal to the force direction. For a so-called *Newtonian fluid* (like water and air), $\sigma_{yx} = \mu \times \partial u/\partial y$. Thus, the unit for μ is

$$[\mu] = \frac{[\sigma_{yx}]}{[\partial u/\partial y]} = \frac{(\text{kgm/s}^2)/\text{m}^2}{(\text{m/s})/\text{m}} = \frac{\text{kg}}{\text{ms}}.$$

It may be noted that μ is, strictly speaking, what is called the *dynamic viscosity* constant. We shall later also meet the *kinematic viscosity*, $\nu = \mu/\rho$. The measure of the wall roughness (e) could, for example, be the typical standard deviation around the mean (think of a cement tube with rough walls). Clearly, the size of e may vary over several orders of magnitude from very smooth glass tubes to steel pipes, cement tubes or even hydropower tunnels in rocks! From Table 1.2, we may derive the dimension matrix

TABLE 1.2
Quantities that may be included in the expression for the pressure drop in the pipe.

Quantity	Name	Unit
Pressure	P	$N/m^2 = kg/s^2 m$
Mean fluid velocity	V	m/s
Tube diameter	D	m
Tube length	L	m
wall roughness	e	m
Viscosity	μ	kg/ms
Density of the fluid	ρ	kg/m^3

	P	V	D	L	e	μ	ρ
kg	1	0	0	0	0	1	1
m	-1	1	1	1	1	-1	-3
s	-2	-1	0	0	0	-1	0

A simple computation shows that the rank is 3, and consequently, we have $7 - 3 = 4$ dimensionless quantities. In this example it is not very smart to use P as a core variable, because we want to express P in terms of the other variables. One possible choice for core variables is $\{V, D, \rho\}$ since

$$\begin{vmatrix} 0 & 0 & 1 \\ 1 & 1 & -3 \\ -1 & 0 & 0 \end{vmatrix} \neq 0,$$

and the columns are thus linearly independent.

The next step is to form dimensionless combinations where the remaining variables are included. It is easy to check that the following combinations are possible choices:

$$\pi_1 = P/(v^2 \rho),$$
$$\pi_2 = L/D,$$
$$\pi_3 = e/D,$$
$$\pi_4 = vD\rho/\mu.$$

(There are other possibilities, but since we are not the first ones to carry out this exercise, we show only the most useful one.) Since we want P to be expressed by the other variables, it is reasonable to think of a relationship of the form

$$\pi_1 = \Phi(\pi_2, \pi_3, \pi_4).$$

Now it is also reasonable to assume (and this is verified by experiments) that the pressure drop is roughly proportional to the tube L. Hence, it should be possible to write

$$\pi_1 = \pi_2 \Phi_2(\pi_3, \pi_4),$$

or

$$P = \frac{L\rho V^2}{D}\Phi_2\left(\frac{e}{D}, \frac{VD\rho}{\mu}\right).$$

In fluid mechanics it is common to replace Φ_2 with $2f_F$, where f_F is called *Fanning's friction factor*,

$$f_F = f_F\left(\frac{e}{D}, \frac{VD\rho}{\mu}\right).$$

The combination e/D is denoted by ε and is known as the tube's *relative roughness*. The second expression,

$$Re = \frac{\rho Dv}{\mu},$$

is the famous *Reynolds number*. Just by applying dimensional analysis, we have established that

$$P = 2\frac{L\rho V^2}{D}f_F(\varepsilon, Re).$$

In the literature one will also encounter the friction factor $f_D = 4f_F$ called *Darcy friction factor*. The friction factor f_F must be determined from more advanced theory and experiments, and in 1944, L.W. Moody presented the famous diagram which is now called a *Moody diagram* (see Figure 1.4).

Note that the friction factor in this diagram is the *Darcy Friction factor*, $f_D = 4f_F$, favored by hydraulic engineers.

For those specially interested, one can mention that for very low Reynolds numbers ($Re < 2000$) the flow will be *laminar*. The laminar flow is the so-called Hagen-Poiseuille flow with a parabolic velocity profile over a smooth tube with circular cross-section. For such flow one can show analytically that $P = 32L\mu V/D^2$, i.e., $f_F = 16/Re$. For the rest of the chart, there are more or less empirical expressions available. When Re is larger than about 6000, $f_F \approx f_C$, where f_C is the solution of *Colebrook equation*

$$\frac{1}{f_C^{1/2}} = -1.74\log\left(\frac{\varepsilon}{3.7} + \frac{1.25}{Re \times f_C^{1/2}}\right).$$

(In the engineering literature, the formula is often stated using $\log_{10}$ instead of the natural logarithm.)

Within the transition between laminar and turbulent flow, the flow is unstable and can switch in an unpredictable way between being laminar and turbulent.

1.3.5   Water waves

In one space dimension, we can write a regular wave on the water surface as

$$\eta(x, t) = a\cos(kx - \omega t),$$

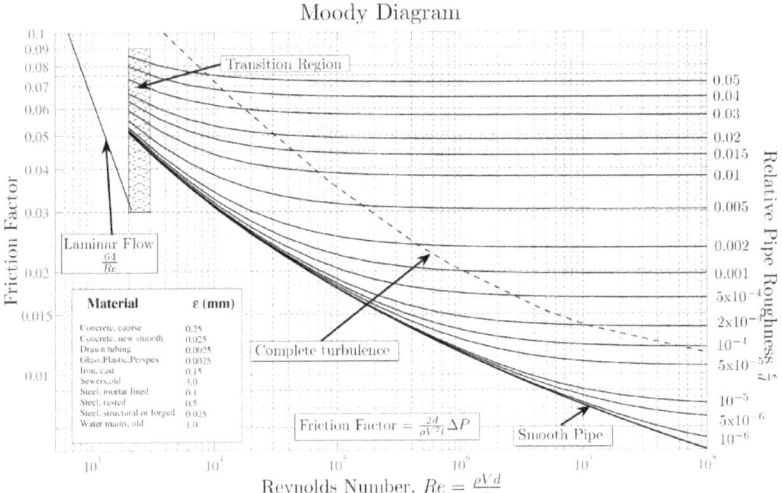

FIGURE 1.4
Moody diagram (Plot created on Matlab by S Beck and R Collins, University of Sheffield, 2008. Wiki Commons (https://commons.wikimedia.org/wiki/File:Moody_diagram.jpg)). Note that the friction factor in this diagram is the *Darcy Friction factor*, $f_D = 4f_F$, favored by hydraulic engineers.

where a is the wave amplitude, x the space coordinate, t is time, $k = 2\pi/\lambda$ is the wavenumber, λ is the wavelength, and ω is the angular frequency (see Figure 1.5). For water waves, k and ω cannot be chosen arbitrarily but must satisfy a *dispersion relation*

$$\omega^2 = f(k, h, a, \cdots).$$

It is reasonable that the angular frequency ω occurs with a second power. Positive and negative frequencies corresponding to waves that move to the right and left, respectively (for positive k). Waves on water may be generated by the wind, boats, etc., and are maintained by gravity. For very short waves, $\lambda = \mathcal{O}(1\text{cm})$, the surface tension keeps the wave going. The surface tension (or stress), σ is characterized by a surface tension coefficient T (not to be confused with the wave period) that connects the surface curvature, defined by the second derivative, and the tension. In one dimension, the expression is $\sigma = T\partial^2\eta/\partial x^2$. The unit for T is thus

$$[T] = \frac{[\sigma]}{[\partial^2\eta/\partial x^2]} = \left(\frac{\text{kgm}}{\text{s}^2}\frac{1}{\text{m}^2}\right)\frac{\text{m}^2}{\text{m}} = \frac{\text{kg}}{\text{s}^2}.$$

Since gravity is important, the gravitational acceleration g and the water density ρ are also possible parameters in the dispersion relation. We neglect the effect of air motion over the waves. Thus, we end up with the following

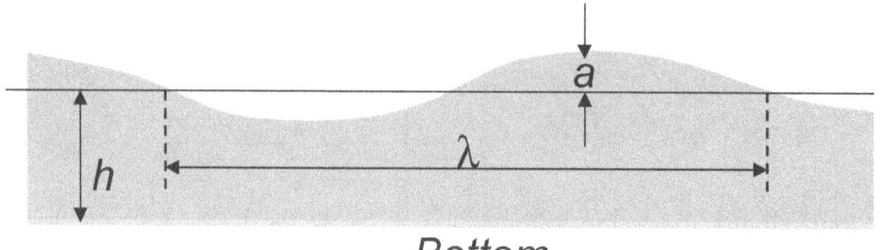

Bottom

FIGURE 1.5
A regular wave on the surface of water.

assumption about the dispersion relation:

$$\omega^2 = f(k, a, h, g, \rho, T).$$

The corresponding dimension matrix is displayed in the following table

	ω^2	k	a	h	g	ρ	T
m	0	-1	1	1	1	-3	0
s	-2	0	0	0	-2	0	-2
kg	0	0	0	0	0	1	1

We easily see that the matrix has rank 3, and consequently, there are $7-3 = 4$ dimensionless combinations. Of several possibilities, we choose $\{k, g, \rho\}$ as our core variables. It is not particularly smart to use ω^2, since we do not want ω^2 to enter on the right side of the equation. Furthermore, both ρ and T (or none) need to be involved since these are the only variables that contain kg in their units. It is now easy to form four dimensionless combinations,

$$\pi_1 = \frac{\omega^2}{kg},$$
$$\pi_2 = ak,$$
$$\pi_3 = hk,$$
$$\pi_4 = \frac{Tk^2}{\rho g},$$

and we find

$$\omega^2 = gk\Phi\left(ak, hk, \frac{Tk^2}{\rho g}\right).$$

In this formula, there are several special cases:

- The wave has very small amplitude compared to the wavelength, $ak \ll 1$.

- The water depth is large relative to the wavelength, $hk \gg 1$.

- The wavelength is much larger than 1cm, $\frac{Tk^2}{\rho g} \ll 1$ (follows from the numerical value of T).

If all three conditions are present, we could write $\omega^2 \approx gk\Phi(0,\infty,0)$. A more refined analysis (by solving the differential equations for water waves) shows that the relation in this case is $\omega^2 \approx gk$ and that $\Phi(0,\infty,0) = 1$. If the depth is not so large, we obtain $\omega^2 = gk\Phi(0,hk,0)$, and a closer analysis here shows that

$$\omega^2 = gk\tanh(hk).$$

If the depth is large and we have very short waves, only surface tension and not gravity is of importance. A simplified dimensional matrix could then be

	ω^2	k	T
m	0	-1	0
s	-2	0	-2
kg	0	0	1

but since only T depends on "kg", it is impossible to combine T with the two others and form a dimensionless combination. Thus, we need another parameter to match "kg", and the only possibility is ρ. We leave it to the reader to show that this gives

$$\omega^2 = C\frac{Tk^3}{\rho} = gkC\frac{Tk^2}{\rho g}.$$

It turns out that also in this case, $C = 1$. In general, it is possible to show by analytical methods that for $ak \ll 1$,

$$\omega^2 = gk\left(\tanh(kh) + \frac{Tk^2}{\rho g}\right).$$

1.3.6 Design of paper aeroplanes

What is the optimal shape of a paper aeroplane? Even if we restrict ourselves to one kind of model, this is not a simple question, and we expect to do a lot of experimentation. Before we proceed, it may be smart to carry out some dimensional analysis. We shall focus on models that have performed well in *Scientific American*'s paper aeroplane competitions. The aeroplanes are made by taking a sheet of paper of length L_0 and width B and fold it with small folds from one side until the centre of gravity lies approximately $1/4$ from the folded edge, as shown in Figure 1.6. After the folding, the plane has length L. Instead of folding, it is also possible, using a somewhat stiffer sheet of paper, to position the centre of gravity correctly by using one or more clips in front of the sheet. Ideally, such a wing should slide with constant velocity U in a fixed angle α with the horizontal plane. Assume that one task is to investigate how the speed depends on the length, width, and weight of the aeroplane. The

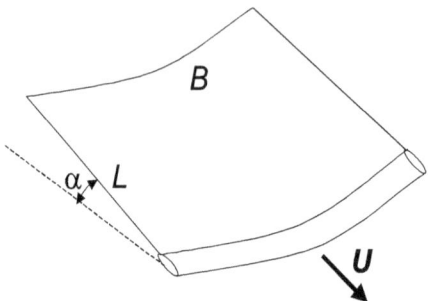

FIGURE 1.6
Sketch of the paper aeroplane.

paper's weight per unit is denoted ρ_p (kg/m^2), the air density ρ_a (kg/m^3), and air kinematic viscosity ν (m^2/s).

Let us first consider the friction force F between the aeroplane and the air. This force must in any case depend on the size and speed of the plane, i.e., L, B, and U. Furthermore, we expect that the viscosity of the air is of importance. If we look at L, B, U, and ν, we find that none of the units of these variables include kg, and since this occurs in force, we need a few more parameters. It is reasonable to choose ρ_a, while there is no reason why ρ_p should enter the expression for the force. There are other parameters that can be expected to have slight impact on the friction force, such as how smooth the paper is, how thick the fold is, etc., but we should already have listed the most important ones. This leads to the following dimension matrix:

	F	B	L	U	ν	ρ_a
kg	1	0	0	0	0	1
m	1	1	1	1	2	-3
s	-2	0	0	-1	-1	0

We leave it to the reader to show that this gives three dimensionless variables which can be arranged so that

$$F = L^2 U^2 \rho_a \Psi \left(\frac{B}{L}, \frac{LU}{\nu} \right).$$

When $B \ll L$, it is reasonable to replace $\Psi \left(\frac{B}{L}, \frac{LU}{\nu} \right)$ with a function of only one unknown, i.e.

$$\Phi \left(\frac{LU}{\nu} \right) = \Psi \left(0, \frac{LU}{\nu} \right).$$

If the aircraft moves with a constant speed at a fixed angle with the horizontal, the friction force has to balance gravity. Then

$$F = Mg \sin \alpha = \rho_p \left(L_0 B \right) \sin \alpha.$$

(In addition, there must be sufficient lift for the aeroplane to stay in the air). For a wide plane, the speed may then be expressed as

$$U^2 = \frac{\rho_p}{\rho_a} \frac{L_0}{L} g \sin \alpha \Phi \left(\frac{LU}{\nu} \right).$$

The combination LU/ν is again the Reynolds number, and as we see, simple dimensional analysis has given us much insight, which we can carry on further in the investigation.

1.3.7 Tritrophic food chain model with general functional response

Consider a food chain consisting of groups of prey X, intermediate predators Y, and top predator Z. It is assumed that top predators feed exclusively on intermediate predators, which themselves consume both prey and alternative food sources. In the absence of intermediate predators, prey populations grow logistically. Similarly, intermediate predators exhibit logistic growth on alternative food when top predators are absent. The model is given by [13]

$$\frac{dX}{dt} = r_X X(1 - X/K_X) - \Phi_X(X) Y,$$

$$\frac{dY}{dt} = r_Y Y(1 - Y/K_Y) + c_1 \Phi_X(X) Y - \Phi_Y(Y) Z,$$

$$\frac{dZ}{dt} = -r_Z Z + c_2 \Phi_Y(Y) Z,$$

where X, Y, Z represent the densities of species (g/m^2), $r_X|r_Y$ the intrinsic growth rates $(/year)$, $K_X|K_Y$ the carrying capacities (g/m^2), r_Z the removal rate of top predators $(/year)$, and $c_1|c_2$ the conversion proportions of biomass (dimensionless). The functional response $\Phi_X(X)$ and $\Phi_Y(Y)$ represent the average density of prey and intermediate predators killed by a single predator $(/year)$, respectively. The dimension matrix provided in the table

	t	X	Y	Z	r_X	r_Y	r_Z	K_X	K_Y	Φ_X	Φ_Y
m	0	-2	-2	-2	0	0	0	-2	-2	0	0
$year$	1	0	0	0	-1	-1	-1	0	0	-1	-1
g	0	1	1	1	0	0	0	1	1	0	0

has a rank of 2. Therefore, the following $(11-2 = 9)$ dimensionless variables

$$\tau = r_X t, \quad x = \frac{X}{K_X}, \quad y = \frac{Y}{K_Y}, \quad z = \frac{r_Y Z}{r_X K_Y}, \quad \varphi_x(x) = \frac{c_1 \Phi_X(X)}{r_X},$$

$$\varphi_y(y) = \frac{\Phi_Y(Y)}{r_Y}, \quad \varepsilon = \frac{c_2 r_Y}{r_X}, \quad \gamma = \frac{r_Z}{r_X}, \quad \kappa = \frac{K_Y}{c_1 K_X}$$

can be used to reduce the system to

$$\frac{dx}{d\tau} = x(1-x) - \kappa\varphi_x(x)\,y,$$

$$\frac{dy}{d\tau} = \varepsilon y(1-y) + \varphi_x(x)\,y - \varphi_y(y)\,z,$$

$$\frac{dz}{d\tau} = -\gamma z + \varepsilon\varphi_y(y)\,z.$$

1.4 Summary

In this chapter we have seen, based on two fairly obvious axioms about nature, that it is possible to derive the quite powerful Buckingham Pi-theorem. These axioms are basically laws of nature, of our universe. The theorem is easy to use, but requires that there really is a relationship between the quantities we have listed. In practice, this can be problematic to determine.

If we were asked to find the eigenfrequency ω of a mathematical pendulum, we would assume that this depends on the length of the pendulum's rod (L), the gravitational acceleration (g), the pendulum's position angle from the vertical at the start (α), and the mass of the bob (m). Based on these quantities, there should exist a relationship

$$\Phi(\omega_0, L, g, \alpha, m) = 0.$$

Here we observe, however, that this is impossible, since the mass is the only quantity containing kg in its unit. Either we must remove m, or there must be an additional quantity that we have forgotten. Since it seems impossible to find other reasonable parameters to include, we are forced to remove m. This leads to the following useful observation:

- *Each fundamental unit must occur in at least two of the quantities.*

The standard procedure now gives us

$$\omega_0 = (g/L)^{1/2} f(\alpha).$$

The results of the dimensional analysis are not unambiguous. Instead of writing

$$\Psi(\pi_1, \pi_2, \cdots, \pi_{M-r}) = 0,$$

we could just as well write

$$\pi_1 = f(\pi_2, \cdots, \pi_{M-r}),$$
$$f(\pi_1, \pi_2) = g(\pi_3, \cdots, \pi_{M-r}),$$

$$\vdots$$

Here, we use what is appropriate. There is no reason to say that one way of writing the formula is more correct than another. In addition, the dimensionless combinations are not unique. If π_1 is dimensionless, then so is also $\sqrt{\pi_1}$, $1/\pi_1$, π_1^2, etc. With more experience, one will often recognize common combinations such as Reynolds number, etc.

The core variables were the subset that we used to form the dimensionless combinations. Usually, there are also several possibilities here. If we are interested in finding how a variable (such as R_1) depends on the others, it is reasonable to avoid using R_1 as one of the core variables. In that way, we find a relation of the form $R_1 = \phi(R_2, R_3, \cdots, R_M)$, that is, R_1 does not enter into the arguments in ϕ.

We have treated dimensional analysis as a method to simplify the relationships between physical quantities. Dimensional analysis is used to obtain an overview and can indicate whether we really understand what we are doing.

One of the best properties of dimension analysis is that it gives us a formulation containing the minimum number of free variables. This is, in particular, valuable for experimental work in the lab or at the computer.

If we decide to find the frequency of a mathematical pendulum by means of experiments only, and assume that $\omega_0 = \Phi(L, g, \alpha, m)$, we may have to determine the function Φ by selecting 10 different values for each variable, that is, perform a total of 10^4 experiments. However, if we first apply dimensional analysis, we realize that it is enough to use only one pendulum, vary the angle α for a reasonable set of values, and then plot α against $\omega_0 \left(L/g\right)^{1/2}$ in order to determine the function $f(\alpha)$ in the expression from the dimensional analysis,

$$\omega_0 = \sqrt{\frac{g}{L}} f(\alpha).$$

A similar simplification is important to do in order to save the number of numerical experiments on a computer, and before setting up experimental plans for statistical experiments.

Dimensional analysis is also crucial when working with *scale models*, that is, doing experiments with models scaled down (or up) in size. Ideally, one would like the dimensionless combinations to be the same for the model as for the original (this is called the *scale laws*).

All comprehensive textbooks on mechanics describe dimensional analysis. For example, both [35] and [30] have nice introductions, while [10] is considered a classic.

1.5 Exercises

1. State the SI-units for the following physical quantities: (*i*) Acceleration, (*ii*) Mass density, (*iii*) Electrical power, (*iv*) Air pressure,

(*v*) Specific heat capacity, (*vi*) The heat conduction coefficient.

2. Mechanical stress has the same unit as pressure (Force per unit area). For a *Newtonian fluid* (like water and air) flowing in the x-direction, the so-called *shear stress* on a plane parallel to the xy-plane is given by

$$\tau = \mu \frac{\partial u(x,y,z)}{\partial y},$$

where u is the velocity in the x-direction at (x,y,z). What is the unit for the constant μ, called the *dynamic viscosity*?

3. Which combinations of *core variables* from the set $\{R_1, \cdots, R_6\}$ may be used if the dimensional matrix is

	R_1	R_2	R_3	R_4	R_5	R_6
F_1	1	1	-1	0	2	2
F_2	-2	-1	1	0	-3	-2
F_3	0	1	0	1	0	2

4. An open cylindrical tank with diameter D is filled to height h with a fluid of density ρ. The bottom has thickness, d, and an elasticity module, E (E is measured in Pascal, like stress). Because of the weight of the fluid, the bottom will sink somewhat, most at the centre (No sinkage at the rims). Show that the sinking (distance, δ) in the centre of the bottom may be expressed as

$$\frac{\delta}{D} = \Phi\left(\frac{h}{D}, \frac{d}{D}, \frac{E}{Dg\rho}\right),$$

where g is the acceleration of gravity.

5. A skydiver in *free fall* with speed U experiences a *drag* (friction force) from the surrounding air. The drag may be written as

$$F_d = \frac{1}{2}\rho_{\text{air}} A U^2 \phi\left(\frac{U\sqrt{A}}{\nu}\right), \tag{1.1}$$

where ρ_{air} is the density of air, A is the cross-sectional area of the skydiver, and ν the kinematic viscosity of the air.

(a) Show how this expression for F_d may be found by dimensional analysis (*Hint*: Use, if necessary, the formula in (1.1) first to determine the units of the involved parameters).

(b) After a while the free-fall jumper will be falling at a constant speed. Find an expression for this speed if we assume that $\phi(x) = 1$. (*Hint*: The force of gravity, pulling the skydiver downwards, is $F_g = mg$, where m is the skydiver's mass and g the acceleration of gravity. Use that F_g is equal to F_d when the speed is constant).

(c) Estimate the free-fall speed in km/hour if we assume that $\phi(x) = 1$.

6. An industrial tank holding a chemical liquid has a hole near the bottom. The chemical is flowing through the hole at an amount Q, measured in m^3/s. It is reasonable to assume that Q depends on the diameter of the hole and the pressure difference Δp in the fluid between the inner and outer sides of the hole. In addition, we expect that the flow is governed by the fluid's density ρ and dynamic viscosity μ. Use dimensional analysis to show that the expression for Q under these assumptions may be written

$$Q = \frac{d^2 \Delta p^{1/2}}{\rho^{1/2}} \phi \left(\frac{d\rho^{1/2} \Delta p^{1/2}}{\mu} \right),$$

where ϕ is an unknown function of only one variable.

7. The force F on an aircraft propeller depends on its diameter d, the speed of the aeroplane U, the density of the air ρ, the number of rotations per second ω, and the viscosity of the air μ. Show how dimensional analysis is used to find the formula

$$F = \rho U^2 d^2 \phi \left(\frac{\omega d}{U}, \frac{Ud}{\mu/\rho} \right), \tag{1.2}$$

where ϕ is an unknown function in two variables (*Hint*: Use, if necessary, (1.2) to find the units for the variables).

8. By measuring the pressure drop p in a tube versus the time t it took to fill a cup with volume V. *Bose* and *Ruert* found around 1910 the relations on Figure 1.7 (left) for water, chloroform, bromoform and mercury. Show, by introducing dimensionless variables (using the density ρ and viscosity μ, $[\mu] = kg\ s^{-1}m^{-1}$), that it exists a common relation covering all the cases. That is, find the variables along the axes in *von Kárman's* alternative presentation of the same data, as shown in the figure to the right.

9. We consider an elastic rubber band which may be stretched many times its original length l_0. The rubber band has a density ρ which we measure in mass per unit length, that is, kg/m. How does the density ρ vary when we stretch the band to a length l from its original length l_0 and density ρ_0? After stretching the band more than twice its original length, we pluck the band like a guitar string. This experiment shows, somewhat unexpected, that the *pitch* (=frequency ω) remains almost constant when we vary the length (*try it!*). However, when stretching the band up towards its breaking limit, the frequency increases somewhat. The force F required for stretching the band to length l is proportional to $l - l_0$ over most

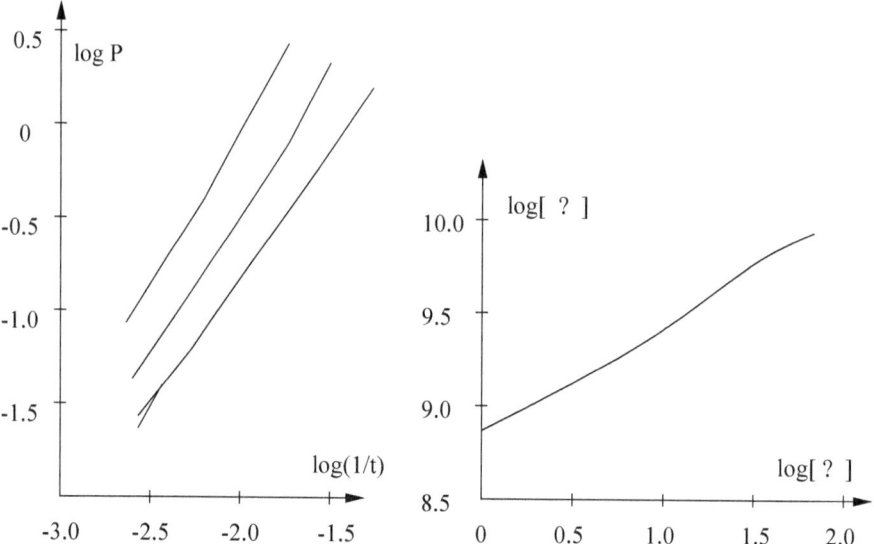

FIGURE 1.7
Presentation of the data in the original paper (left) and vonKarman's revised
graph after applying dimensional analysis (right).

of the range, that is, $F = F_0 \frac{l-l_0}{l_0}$, where F_0 is a constant. Use dimensional analysis to explain the behaviour of the frequency. (*Hint:*
Assume first that $\omega = f(l, \rho, F)$, apply dimensional analysis, and
then introduce the expression of the density as a function of l, l_0
and ρ_0).

10. In forest assessment one wants to estimate the volume V (also called
the *cubic content*) of a tree by measuring its height (h) and diameter
(d) at the root. A test example in the documentation of *Minitab®*
suggests the following formula (based on multilinear regression) for
American cherry trees:

$$V^{1/3} = \beta_0 + \beta_1 d + \beta_2 h + \beta_3 d^2. \qquad (1.3)$$

Here $\{\beta_i\}$ are regression coefficients calculated from a set of calibration data.

(a) Americans use *foot* and most of the rest of the world *meter* as
the basic length unit. Is it possible to use the same values for
$\{\beta_i\}$ in both cases?

(b) Show that dimensional analysis, based on the variables v, d,
and h, instead recommends applying a relationship of the form

$$\pi_1 = \phi(\pi_2). \qquad (1.4)$$

Find π_1 and π_2, and give examples of what the function ϕ could be for some "idealized trees", e.g., cylinders and cones.

(c) For students who know *Minitab*, it might be interesting to check which of the models is best: The model in (1.3), or a regression model based on (1.4). The calibration data used in Minitab 14 are stored in the MTW file trees.mtw.

11. The necessary force (F) to keep a ship at a constant speed (U) depends on its shape; primarily the length (L), width (W), and its depth into the water (D). In addition, the water density, ρ, the viscosity, ν, and the acceleration of gravity, g, play a part. Use dimensional analysis to find an expression for the force which includes the two most famous dimensionless numbers in ship design:

$$\text{Froude number: } Fr = U/\sqrt{Lg},$$
$$\text{Reynolds number: } Re = LU/\nu.$$

Ideally, a scale model[1] of the ship should be tested experimentally in water by keeping the dimensionless numbers for the model equal to those of the original ship. Is this really possible?

(*Hints*: $[F] = \text{kgm/s}^2$, $[\rho] = \text{kg/m}^3$, $[\nu] = \text{m}^2/\text{s}$, $[g] = \text{m/s}^2$).

---

[1] A *scale model* is a model of the ship with the same geometric shape, but with a smaller size (Say, $L = 1$m for the model, compared to 200m for the original ship).

2

Scaling

2.1 Introducing Scaled Variables

After establishing a mathematical model in the form of an equation, it will be necessary to introduce dimensionless variables. Usually, it is not difficult to do this, but it can be carried out in several ways, and it is not always easy to see what is the most appropriate way. However, there exists an intelligent way of doing this called *scaling the equations*. When the equations are scaled, it is easy to see which parts are important and which are less important. It may be difficult to scale equations, and in any case this will depend on the problem we are considering, even if the equation is basically the same all the time. Somewhat simplified, we can say that the scales force us to think about the situation, and in this way, we gain insight into what we are doing. The theory in this chapter is mainly taken from [30].

- *To scale a variable u^* means to write the variable as*

$$u^* = Uu,$$

where $[U] = [u^*]$, U is of the same order of magnitude as u^*, and u is of order 1.

Here U is the characteristic size of u^*. If we use U as our unit of measurement, u is neither particularly large nor particularly small. This is a somewhat imprecise definition, but it reflects the fact that scales are not always very well defined.

Until getting used to scaling variables, it is handy to have a notation in order to distinguish between the original variables with units and the new dimensionless variables. We will do this, as suggested in [30], by attaching * on the original variables, and remove * after the variable has been made dimensionless. After a while, we become tired of writing *, and understand the transition from the context.

Let us consider a variable u^* which is a function of time t^*. It is usually reasonable to use

$$U = \max_{t^*} |u^*(t^*)|$$

DOI: 10.1201/9781003725206-2

as a scale for u^*, even if the minimum of $|u^*|$ is much smaller. Then, at least $|u|$ is less than or equal to 1. In practice, this often means to *estimate* the maximum value, since we may not know u^* in detail.

It will also be necessary to find scales for time. Sometimes the maximum value of t^* may be used, but more often the scale is defined as a period over which u^* *varies significantly*. If $u^*(t^*) = \sin at^*$, a reasonable time scale would be $1/a$, since u^* then varies from 0 to $\sin 1$. As suggested in [30], it is often possible to find a reasonable time scale by looking at (or estimate)

$$\frac{\max |u^*(t^*)|}{\max |du^*/dt^*|}.$$

Such expressions must be used with common sense, and when working with scales, we are not very careful about extra factors such as 2, π, etc. Scaling is not an exact science – often a rough estimate is all we need.

- *To scale an equation means to introduce dimensionless variables based on the scales of the variables in the equation.*

Depending on the situation we are in, the same equation could be scaled in several ways. After the equation is scaled, it will be clear what are important and less important parts of the equation (if not all parts are equally important). Often one will be able to get approximate solutions by solving the equation when the less important parts are neglected. Knowing the scales of the variables of a mathematical model requires knowledge and physical understanding, and is one of the most important things we do in mathematical modelling. As will be seen below, scaling is not nearly as easy as it sounds. A good example is one of the main modelling examples in [30], where the authors, several years after the book was published, discovered that the time scale they had suggested was not really appropriate (recently, the authors of these notes have published a completely different scaling of the same equations).

2.2 Order of Magnitude

We say that the function $f(x)$ is of the *order of magnitude* $g(x)$ when $x \to a$ if there exist two finite numbers $\{m, M\}$, $0 < m < 1 < M$, such that

$$m \leq \frac{f(x)}{g(x)} \leq M$$

for $x \to a$. This is written

$$f(x) = \mathcal{O}(g(x)), \quad x \to a,$$

and expressed in words as "$f(x)$ *is of order* $g(x)$ *when* x *is approaching* a". Some like to require that $m = \sqrt{1/10}$ and $M = \sqrt{10}$ (what is then $\log_{10} m$ and $\log_{10} M$?), but we prefer a more informal use, that is,

$$\log(1+x) - x = \mathcal{O}(x^2) \text{ for small } x\text{-s.}$$

For series, the first non-zero term is called the *leading order* term. For example, $4x^3 + 3x^4 + 5x^5 + \cdots$ is *of leading order* x^3 for small x-s.

A slightly different symbol, "$o\,()$" is more precise: We write

$$f(x) = o(g(x)) \text{ when } x \to a,$$

if

$$\lim_{x \to a} \frac{f(x)}{g(x)} = 0.$$

Thus,

$$\sin x - x + x^3/6 = \mathcal{O}(x^5),$$
$$\sin x - x + x^3/6 = o(x^4),$$

when $x \to 0$.

2.3   A Simple Case Study

In the following artificial and simple example, we shall see how scales change depending on the nature of the problem. The example is trivial and easy to solve analytically. The assumption about the friction force is not very realistic.

A spherical ball is fired vertically into a viscous fluid as illustrated in Figure 2.1. The ball's initial speed is V, and the forces acting on the ball are

$$
\begin{aligned}
Gravity: & \quad gm \\
Friction: & \quad -k\frac{\mathrm{d}x^*}{\mathrm{d}t^*} \\
Buoyancy: & \quad -gm\frac{\rho_v}{\rho_k}
\end{aligned}
$$

(Here, ρ_v is the fluid density and ρ_k the density of the ball). The equation of motion follows from Newton's Law, and we assume that the ball starts at $x^* = 0$ with velocity V:

$$m\frac{\mathrm{d}^2 x^*}{\mathrm{d}t^{*2}} = gm - k\frac{\mathrm{d}x^*}{\mathrm{d}t^*} - mg\frac{\rho_v}{\rho_k},$$

$$x^*(0) = 0, \quad \frac{\mathrm{d}x^*}{\mathrm{d}t^*}(0) = V.$$

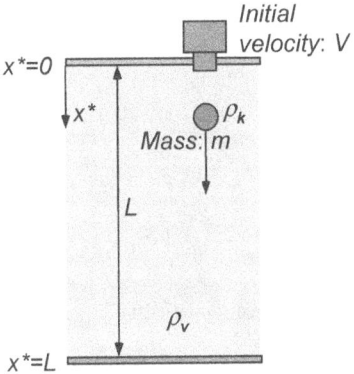

FIGURE 2.1
Ball falling or fired into a viscous fluid.

We shall also assume that $\rho_v < \rho_k$, such that the ball does not eventually return to the surface, and we replace $g(1 - \frac{\rho_v}{\rho_k})$ with a modified g so that the problem simplifies to

$$m\frac{\mathrm{d}^2x^*}{\mathrm{d}t^{*2}} = gm - k\frac{\mathrm{d}x^*}{\mathrm{d}t^*},$$

$$x^*(0) = 0, \quad \frac{\mathrm{d}x^*}{\mathrm{d}t^*}(0) = V.$$

In this case we can imagine a number of special cases. If the ball had fallen freely with zero initial velocity, it would, at $x = L$, have reached the speed v_{FF} where $v_{FF} = \sqrt{2Lg}$ (Vertical motion under constant acceleration). If, on the other hand, the medium is *very* viscous (think of syrup!), the ball will, after a while fall with constant speed v_0 determined by

$$0 = gm - kv_0,$$

i.e. $v_0 = \frac{gm}{k}$.

   Below we shall consider three different situations, and it will turn out that the ratio between v_{FF} and v_0 is crucial.

2.3.1 Case A: Large friction – what happens initially is not important

This is a situation where either L or the viscosity (here expressed by the constant k) is so great that the ball falls at a constant speed over most of the distance. Assuming that the ball has speed v_0 all the way, we may estimate the time it takes to go from $x^* = 0$ to $x^* = L$ to about

$$T_0 = \frac{L}{v_0} = \frac{Lk}{mg},$$

and this gives us a reasonable time scale. Depending on the size of V, the actual time the ball uses could be slightly larger or smaller than T_0. There is, however, an implicit assumption here that V is not very large compared to v_0. The length scale is not a problem, we use L and introduce dimensionless variables x and t as

$$x^* = Lx,$$

$$t^* = \frac{Lk}{mg}t.$$

By bringing this into the equations, we obtain

$$m\frac{\mathrm{d}^2\left(Lx\right)}{\mathrm{d}\left(\frac{Lk}{mg}t\right)^2} + k\frac{\mathrm{d}\left(Lx\right)}{\mathrm{d}\left(\frac{Lk}{mg}t\right)} = gm,$$

$$Lx\left(0\right) = 0, \ \frac{\mathrm{d}\left(Lx\right)}{\mathrm{d}\left(\frac{Lk}{mg}t\right)}\left(0\right) = V,$$

and after simplification,

$$\frac{gm^2}{Lk^2}\frac{\mathrm{d}^2x}{\mathrm{d}t^2} + \frac{\mathrm{d}x}{\mathrm{d}t} = 1,$$

$$x(0) = 0, \ \frac{\mathrm{d}x}{\mathrm{d}t}(0) = \frac{V}{v_0}.$$

In addition to the variables x and t, the problem contains two dimensionless parameters:

$$\varepsilon = \frac{gm^2}{Lk^2},$$

$$\mu = V/v_0.$$

We note that

$$\varepsilon = \frac{gm^2}{Lk^2} = 2\frac{1}{2Lg}\left(\frac{gm}{k}\right)^2 = 2\frac{v_0^2}{v_{FF}^2} = 2\left(\frac{v_0}{v_{FF}}\right)^2.$$

Thus, ε is a small parameter (compared to 1) if $v_0 \ll v_{FF}$. It is characteristic for this case that the speed v_0 is much less than the speed the ball would have had at $x^* = L$ if it fell freely. It is typical that when we have scaled the equations, the dimensionless parameters have interesting interpretations that we may apply for *hindsight*.

After the scaling is complete, the equation has the form

$$\varepsilon\frac{\mathrm{d}^2x}{\mathrm{d}t^2} + \frac{\mathrm{d}x}{\mathrm{d}t} = 1, \ x(0) = 0, \ \frac{\mathrm{d}x}{\mathrm{d}t}(0) = \mu,$$

with $\varepsilon = 2\frac{v_0^2}{v_{FF}^2}$, $\mu = \frac{V}{v_0}$.

As mentioned above, there is here an assumption that μ is not particularly large. In that case, one might imagine another time scale (see Case C below).

2.3.2 Case B: Small friction – nearly free fall

This problem could have been the same as in Case A, but now with the difference that L is so small that the ball never reaches speeds near to v_0. Thus, friction is of little importance.

Again, L is a natural length scale for the x^*. If the ball fell freely and $V = 0$, the ball would fall with nearly constant acceleration, and the time it takes to fall to $x^* = L$ would roughly be $\sqrt{2L/g}$. Since we have already introduced v_{FF}, we apply $T_0 = 2L/v_{FF}$ as our scale. Certainly, T_0 is only about the half of $\sqrt{2L/g}$, but we do not care about this for a scale estimate. We have already assumed that the speed V is so small that it does not affect the time scale. With these deliberations, we may write

$$x^* = Lx,$$

$$t^* = \frac{2L}{v_{FF}}t,$$

and obtain

$$mL\frac{2Lg}{4L^2}\frac{\mathrm{d}^2x}{\mathrm{d}t^2} + kL\frac{v_{FF}}{2L}\frac{\mathrm{d}x}{\mathrm{d}t} = mg, \quad x(0) = 0, \quad \frac{\mathrm{d}x}{\mathrm{d}t}(0) = \frac{2V}{v_{FF}},$$

and finally

$$\frac{1}{2}\frac{\mathrm{d}^2x}{\mathrm{d}t^2} + \varepsilon\frac{\mathrm{d}x}{\mathrm{d}t} = 1, \quad x(0) = 0, \quad \frac{\mathrm{d}x}{\mathrm{d}t}(0) = \mu,$$

$$\varepsilon = \frac{v_{FF}}{2v_0}, \quad \mu = \frac{2V}{v_{FF}}.$$

Note that the definition of ε has changed compared to Case A, and here, ε is a small parameter if $v_{FF} \ll v_0$. This is a characteristic feature of Case B. The scaling above is only reasonable if V is small compared to v_{FF}. If V is greater than v_{FF}, but still smaller than v_0, the ratio L/V could be a reasonable time scale. We leave to the reader to complete the scaling in this case.

2.3.3 Case C: Ball released into a highly viscous medium. Initial velocity much larger than v_0

In this case, we expect that friction dominates over gravity, and we estimate the length and time scales by looking at the approximate equation

$$m\frac{\mathrm{d}^2x^*}{\mathrm{d}t^{*2}} = -kV, \quad x^*(0) = 0, \quad \frac{\mathrm{d}x^*}{\mathrm{d}t^*}(0) = V.$$

If this were the exact equation, the ball would stop for $t^* = T_0 = \frac{m}{k}$ (since $\frac{\mathrm{d}x^*}{\mathrm{d}t^*} = V - \frac{Vk}{m}t^*$). An associated length scale (where we again disregard a factor of 2) will then be

$$L = VT_0 = \frac{Vm}{k}.$$

TABLE 2.1

A summary of the scaling example

Case	Characteristics	Length scale	Time scale	Equation	Parameters
A	$v_0 \ll v_{FF}$	L	L/v_0	$\varepsilon\ddot{x} + \dot{x} = 1$	$\varepsilon = 2\frac{v_0^2}{v_{FF}^2},\ \mu = \frac{V}{v_0}$
B	$v_0 \gg v_{FF},$ $V < v_{FF}$	L	L/v_{FF}	$2\ddot{x} + \varepsilon\dot{x} = 1$	$\varepsilon = \frac{v_{FF}}{v_0},\ \mu = v/v_{FF}$
C	$V \gg v_0$	mv/k	m/k	$\ddot{x} + \dot{x} = \varepsilon$	$\varepsilon = \frac{v_0}{v},\ \mu = 1$

We introduce $x^* = \frac{mV}{k}x$ and $t^* = \frac{m}{k}t$:

$$m\frac{mV}{k}\frac{k^2}{m^2}\frac{\mathrm{d}^2 x}{\mathrm{d}t^2} + k\frac{mV}{k}\frac{k}{m}\frac{\mathrm{d}x}{\mathrm{d}t} = mg, \quad x(0) = 0, \quad \frac{\mathrm{d}x}{\mathrm{d}t}(0) = 1,$$

which, after some simplification, becomes

$$\frac{\mathrm{d}^2 x}{\mathrm{d}t^2} + \frac{\mathrm{d}x}{\mathrm{d}t} = \varepsilon, \quad x(0) = 0, \quad \frac{\mathrm{d}x}{\mathrm{d}t}(0) = 1,$$

with $\varepsilon = \frac{v_0}{V}$.

In this situation, ε is a small parameter when $V \gg v_0$, and this is the characteristic feature for Case C.

2.3.4 Summary

We have now seen three different situations where weight has been put on various parts of the equation. The problem has, in addition to V, two characteristic speeds, namely $v_{FF} = \sqrt{2gL}$ and $v_0 = \frac{gm}{k}$, and the various situations above are characterized by the mutual size of these speeds. We summarize the results in Table 2.1.

In all three situations we end up with a parameter ε which is typically small. The related terms in the equation are also small, and by neglecting the terms of order ε, we obtain the simplified equations.

Although it is the rule rather than the exception that we end with terms of different size in a scaled equation, it is also possible that all terms happen to be of the same magnitude.

We leave it to the reader to show that the exact solution is

$$x^*(t^*) = \frac{gm}{k}t^* + \left(V - \frac{gm}{k}\right)\frac{m}{k}\left(1 - e^{-t^*/(m/k)}\right),$$

and the graph in Figure 2.2 shows how the exact solutions relate to the situations we have seen. Note that the cases we have considered by no means cover the entire chart. By setting $\varepsilon = 0$ for all three situations above, we obtain simplified equations, but Case A is special. With $\varepsilon = 0$, the equation and

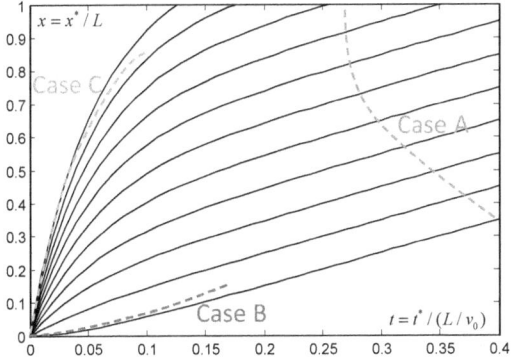

FIGURE 2.2
The figure shows the exact paths and an indication of the three situations we
have considered.

initial conditions become

$$\dot{x}_0 = 1, \tag{2.1}$$

$$x_0(0) = 0, \tag{2.2}$$

$$\dot{x}_0(0) = \mu, \tag{2.3}$$

and unless μ happens to be 1, it is *impossible* to solve the simplified problem
exactly. The general solution to Eq. (2.1) is

$$x_0(t) = C + t,$$

and since the speed is 1 (v_0 in the original variables), the form is reasonable.
The approximate solution is simply not valid near 0, and in order to determine the correct $C = C(\mu)$ a special technique (*singular perturbation*) will be
required.

In Case B, the approximate equation is

$$2\ddot{x}_0 = 1,$$

$$x_0(0) = 0,$$

$$\dot{x}_0(0) = \mu,$$

which we immediately solve as

$$x_0(t) = t^2/4 + \mu t.$$

We can check the approximate solution by inserting it in the exact equation,

$$2\ddot{x}_0 + \varepsilon\dot{x}_0 - 1 = \varepsilon\left(\frac{t}{2} + \mu\right).$$

The error on the RHS increases with time, and this is reasonable since the approximate solution is not at all limited by friction.

The equation for Case C has the approximate solution

$$x_0(t) = 1 - e^{-t},$$

so here $x_0(t) < 1$ for all t.

Although the solutions in B and C obviously have their weaknesses, they are great for the situations they are supposed to cover. Convince yourself by drawing the approximate and exact solutions for some choice of ε and μ.

2.4   Scaling Considerations

Arguments based on scale considerations have proven to be quite useful in many contexts, but they require some physical insight and creativity and are not always so easy to perform.

2.4.1   Turbulence

Fluids are mixed (on the microscopic level) by *molecular diffusion*, and (on the macroscopic level) by *convection*. Molecular diffusion is related to the kinematic viscosity of the fluid (ν, $[\nu] = \mathrm{m}^2/\mathrm{s}$), while convection is the macroscopic motion, typically visible whirls, observed when we move the spoon around in a cup of tea or when watching the whirling water in a river, for example.

Suppose we consider a whirl with diameter L. The time scale associated with L and convection with a velocity scale U will be

$$t_K = L/U.$$

The time scale associated with diffusion over a length L may likewise be expressed by the kinematic viscosity, ν and L. The only possibility is

$$t_D = \frac{L^2}{\nu}.$$

We observe that the quotient between these two scales is

$$\frac{t_D}{t_K} = \frac{L^2 U}{\nu L} = \frac{LU}{\nu} = Re,$$

which is a new meaning of the well-known *Reynolds number*, also mentioned above. A Reynolds number $Re \ll 1$ indicates that the mixing is dominated by molecular diffusion, whereas $Re \gg 1$ means that it is dominated by convection.

The value of ν for water is about $10^{-6}\mathrm{m}^2\mathrm{s}^{-1}$. Consider a typical river with width $L = 100\mathrm{m}$, and $U = 1\mathrm{m}/\mathrm{s}$. Then

$$Re \approx \frac{100 \cdot 1}{10^{-6}} = 10^8,$$

and the mixing of the water in the river is entirely dominated by convection.

In turbulent flow large vortices initiate motion of small vortices which, in turn, set into motion (and keep alive) even smaller vortices, and so on. In the very small vortices, viscosity will reduce the motion, and the kinetic energy is eventually transferred into heat. The kinetic energy dissipation (loss of energy) is mainly from these small vortices with a length scale l' and velocity scales u'. We can estimate the energy loss (E) per time and unit mass by assuming that $E = E(l', u', \nu)$, and that $E \propto u'^2$ (in other words, proportional to the kinetic energy that is present in the smallest vortices). Simple dimensional analysis then gives

$$E \propto \nu \left(\frac{u'}{l'}\right)^2,$$

and an estimate for E will be $E = \nu\left(\frac{u'}{l'}\right)^2$. From the above we can further assume that the smallest vortices have $Re \approx 1$, or $t_K \approx t_D$, that is

$$\frac{l'u'}{\nu} = 1.$$

Thus,

$$l' = \left(\frac{\nu^3}{E}\right)^{1/4},$$

$$u' = (\nu E)^{1/4}.$$

These scales are called the *Kolmogorov's micro scales* in turbulence theory. These are the smallest scales that occur before the diffusion takes over and turns the kinetic energy over to heat by internal friction.

If we mix 1kg of water with a mixer with an output of 100W, this power would disappear in the smallest vortices, and consequently the diameter of these vortices is of the order

$$l' = \left(\left(\frac{10^{-6}\mathrm{m}^2}{\mathrm{s}}\right)^3 \Big/ \left(100\frac{\mathrm{kgm}^2}{\mathrm{s}^2 \cdot \mathrm{s} \cdot 1\mathrm{kg}}\right)\right)^{1/4} = 10^{-5}\mathrm{m} = 0.01\mathrm{mm}.$$

2.4.2 Geometric similarity of animals

Why do we look like we do? It has long been known that animal forms are not just random, but a result of the strength of muscles and bones in relation to the strength of gravity here on Earth. If we could reduce a human to Thumbelina-size, it turns out that the body would immediately be torn to pieces by the muscles. Therefore, insects usually have very small muscles (thin legs!) in relation to their size.

The discussion below is taken from the unpublished note *Dimensional Analysis* by Professor Kristian B. Dysthe, University of Tromsø, 1992. One of his references is the world famous book *On Growth and Form* by D'Arcy W. Thompson, first published in 1917 [40].

We shall first look at animals approximately geometrically similar, having a typical length scale L. We may then argue that their

i. weight is proportional to their volume, that is, $\propto L^3$,

ii. muscle power is proportional to the amount of muscle fibres, which in turn are proportional to the muscle cross-sectional area, $\propto L^2$,

iii. ability to do work (and produce heat), power $\propto$ lung capacity $\propto$ oxygen uptake $\propto$ surface of the lungs $\propto L^2$ (may be somewhat questionable because of the fractal structure of the lungs).

2.4.3 Jumping

When an animal wants to jump into the air, it must produce a certain amount of energy which becomes its kinetic energy the moment it leaves the ground. The energy is produced by accelerating the body over a distance, $\mathcal{O}(L)$, multiplied by the power it generates, $\mathcal{O}(L^2)$. In other words, the supply of kinetic energy = force × distance $\propto \mathcal{O}(L^2) \times \mathcal{O}(L) = \mathcal{O}(L^3)$. The necessary potential energy for a jump of height H will likewise be $H = Hmg \propto HL^3$, where m is the mass of the animal. This therefore gives

$$HL^3 = \text{const.} \cdot L^3,$$

or that H is constant. Thus, we get the somewhat surprising result that all animals of the same shape jump equally high!

It should also be similar to jump down a certain height H and land in a controlled manner. Cats seem to have this property, and otherwise, it is alleged that the kangaroo and a jumping mouse (which by their names would have slightly different size!) can jump equally high.

2.4.4 Running uphill

From the observation above, the power an animal manages to maintain will be proportional to L^2, and since the required power to keep a speed v up a hill with a slope angle α is

$$(mg \sin \alpha) \cdot v \propto L^3 v,$$

we obtain

$$v L^3 = \text{const.} \times L^2,$$

or
$$v \propto 1/L.$$

Small animals can therefore keep a higher speed than large animals when running uphill.

2.4.5 Diving animals

Assume that all animals during a dive are moving at speed v. The friction force that must be overcome will typically be proportional to the square of the velocity and cross sectional area of the animal, i.e., $F \propto v^2 L^2$ (This may be concluded from formulas derived from dimensional analysis). The total energy consumed within the water is $F \cdot (v \cdot t_{max})$, where t_{max} denotes the maximum time it can stay under water. Since the energy stored in the animal will be proportional to L^3, then

$$L^2 \cdot t_{max} \propto L^3,$$

or
$$t_{max} \propto L.$$

This means, in other words, that large animals can stay longer under water than small animals, and this we know from the marine mammals.

We leave it to the reader to speculate about any other scaling arguments, for example, what the Mars and Jupiter residents would look like.

Finally, we shall consider an example from sport.

2.4.6 Weightlifting

For equally shaped weight lifters, the muscle strength is proportional to L^2, and the weight is proportional to L^3. In other words, the force should be proportional to the weight of power $2/3$. Figure 2.3 shows that this holds astonishingly well.

2.5 Exercises

1. The following expressions have been proposed as the time scale for the function $u^* (t^*) = A \cos (2\pi f_0 t^*)$:

$$T = 1/f_0,$$
$$T = 1/ (2\pi f_0),$$
$$T = 1/ (\pi f_0),$$
$$T = 500\pi f_0^{-1}.$$

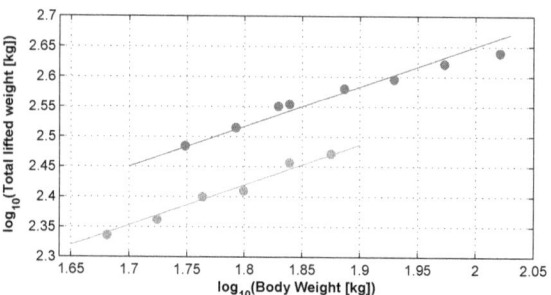

FIGURE 2.3

The world records in Snatch + Jerk + Press as a function of the lifter's body weight (women red dots, men blue dots). Note that the line has slope 2/3

May all these be used as the time scale?

2. A common mathematical model for the size of a population $y^*\left(t^*\right)$ as a function of time t^* is described by the *logistic equation*

$$\frac{dy^*}{dt^*} = ry^*\left(1 - \frac{y^*}{K}\right).$$

Here r is called the *growth rate* and K the *sustainable capacity*.

(a) Which scale is suitable for y^*?

(b) Determine a time scale when $y^* \ll K$.

(c) Introduce these scales into the equation so that it becomes dimensionless (The equation can easily be solved by inserting $y = 1/u$ and solving for u).

3. In many dynamic systems one talks about *time constants*. For an exponential function, $u\left(t\right) = A\exp\left(-at\right)$, the *time constant* is defined as follows: First draw the tangent to $u\left(t\right)$ at t_0. This tangent is crossing the x-axis at t_1, and the time constant is defined $T = \left|t_1 - t_0\right|$. Show that this definition also follows from the Lin&Segel *Rule of Thumb*,

$$T = \frac{\max\left|u\left(t\right)\right|}{\max\left|du\left(t\right)/dt\right|}.$$

3

Perturbation Analysis

3.1 Regular Perturbation

In this section we shall consider a way to handle equations containing small parameters, and the scaled equations from Case B and Case C in the case study in Sec. 2.3 are of this form. The basic idea is to write the solution as a power series in the small parameter and determine the terms in the series recursively. We shall take a closer look at this methodology and show how it works on some simple examples. Regular perturbation is one of the most common techniques in traditional applied mathematics, and is well treated in several textbooks. The presentation below is incomplete, but adapted to what we are going to need.

We have a *regular perturbation problem* if we have an equation

$$D(x, \cdots, \varepsilon) = 0,$$

containing a small parameter ε so that the full solution, x_{sol}, approaches solution x_0 of the reduced equation

$$D(x, \cdots, 0) = 0,$$

when ε tends to 0. The statement is very imprecise, as we say nothing about how x_{sol} approaches x_0. If we then know that ε is small (after the equation is scaled), we may approximate the complete solution x_{sol} with x_0. This is pursued further by writing the solution in the form of a power series in ε,

$$x_{sol} = x_0 + x_1\varepsilon + x_2\varepsilon^2 + x_3\varepsilon^3 + \cdots,$$

for then to come up with a sequence of simpler equations for x_0, x_1, $\cdots$. Since ε is small, we expect that the terms in the series become smaller and smaller, and that the approximation gets better the more terms we include. In practice, it is not that easy. The solution of the equations for x_i often gets more complicated as i increases, and the power series do not tend to have very impressive convergence properties.

If we forget about these objections, the method of regular perturbation is easy to state:

DOI: 10.1201/9781003725206-3

1. Write the solution as a power series in ε,

$$x_{sol} = x_0 + x_1\varepsilon + x_2\varepsilon^2 + x_3\varepsilon^2 + \cdots$$

2. Put the series into the equation and clean up the expression so that we obtain a new power series in ε,

$$D(x_{sol}, \varepsilon) = D(x_0 + x_1\varepsilon + x_2\varepsilon^2 + x_3\varepsilon^2 + \cdots, \varepsilon),$$
$$= P(x_0, 0) + P_1(x_0, x_1)\varepsilon + P_2(x_0, x_1, x_2)\varepsilon^2 + \cdots$$

3. Set each coefficient in the series equal to 0 and solve the equations to get *recursively*:

$$P_0(x_0, 0) = 0,$$
$$P_1(x_0, x_1) = 0,$$
$$P_2(x_0, x_1, x_2) = 0,$$
$$\vdots$$

This method gives us x_0, x_1, x_2, $\cdots$, and the idea may be used in many connections:

- For approximate solutions to algebraic and transcendental equations.

- For approximate expressions to integrals.

- For ordinary and partial differential equations.

Perturbation analysis is often complementary to numerical techniques. In many situations, numerical methods have problems when ε is small (this is especially the case for *singular perturbation* discussed later). The perturbation analysis gives us the asymptotic relations which are useful when ε goes to 0, in contrast to a small number of numerical calculations where we need to keep ε fixed for each calculation. In other contexts there is no really small (or large) parameter to use, and there is no way around numerical calculations.

Perturbation analysis had its best days before we had computers with opportunities for large scale numerical calculations. In particular, in the field of aerodynamics and other fluid mechanics, perturbation analysis has been widely recognized. Today, there are computer programs for symbolic manipulation enabling us to find perturbation solutions of orders we could only dream about. However, sometimes profit is marginal; if one does not achieve reasonable approximations with one or two terms, there is often little to gain by calculating more terms.

3.1.1 The projectile problem

The Projectile Problem, discussed in the book of Lin & Segel [30], pp. 233, is a simple and instructive example of how regular perturbation works. The problem leads to a non-linear differential equation where it is not possible to write the solution in explicit form using elementary functions.

3.1.1.1 The model

A projectile is sent vertically up from a planet without an atmosphere. The motion is described by the position $x^*(t^*)$ above the planet's surface, where t^* is the time and

$$x^*(0) = 0, \quad \frac{dx^*}{dt^*}(0) = V. \tag{3.1}$$

The projectile will be affected by a force given by Newton's law of gravitation, that is,

$$F(x^*) = -G\frac{Mm}{(R+x^*)^2},$$

where G is the gravitational constant, M is the planet's mass, R the planet's radius, and m the projectile's mass. Similar to Earth, the gravity force on the planet's surface may be written as $F(0) = -mg$, so that $g = GM/R^2$. Thus, it follows that

$$m\frac{d^2x^*}{dt^{*2}} = -\frac{R^2gm}{(R+x^*)^2}. \tag{3.2}$$

The mathematical model thus consists of the non-linear differential equation

$$\frac{d^2x^*}{dt^{*2}} = -\frac{R^2g}{(R+x^*)^2},$$

with the initial conditions stated in (3.1).

3.1.1.2 Scaling

We are going to study a situation where V is much smaller than the planet's *escape velocity*. If V is larger than the escape velocity, the projective will leave the planet permanently. For Earth the escape velocity is about 11.2km/s. However, here the assumption implies that $x^*(t^*) \ll R$ for the whole trip of the projectile. Under this assumption, the equation simplifies to

$$\frac{d^2x^*}{dt^{*2}} = -\frac{R^2g}{(R+x^*)^2} = -\frac{g}{(1+x^*/R)^2} \approx -g.$$

This equation may easily be solved with the given initial conditions:

$$x^*(t^*) \approx -\frac{1}{2}gt^{*2} + Vt^*. \tag{3.3}$$

The approximate maximum height follows from (3.3) by observing that the time to maximum height is approximately given by

$$\frac{dx^*}{dt^*} \approx -gt^* + V = 0,$$

or

$$t_{\max} \approx \frac{V}{g}.$$

Thus,

$$x_{\max} \approx -\frac{1}{2}g\left(\frac{V}{g}\right)^2 + V\left(\frac{V}{g}\right) = \frac{1}{2}\frac{V^2}{g}.$$

Reasonable scales for (3.2), where we do not care about factors of 2, will now be

$$X = \frac{V^2}{g}, \ T = \frac{V}{g}.$$

Inserted into the equation, this leads to

$$\frac{d^2\left(\frac{V^2}{g}x\right)}{d\left(\frac{V}{g}t\right)^2} = -\frac{R^2 g}{\left(R + \frac{V^2}{g}x\right)^2},$$

$$\frac{V^2}{g}x(0) = 0, \ \frac{d\left(\frac{V^2}{g}x\right)}{d\left(\frac{V}{g}t\right)} = V,$$

and after cleaning up, we have the scaled problem,

$$\ddot{x} = -\frac{1}{(1+\varepsilon x)^2}, \tag{3.4}$$

$$x(0) = 0, \ \dot{x}(0) = 1, \ \varepsilon = \frac{V^2}{Rg}. \tag{3.5}$$

It turns out that it is not possible to express the solution of this equation, $x = x(t, \varepsilon)$, by means of elementary functions (this is not quite obvious!). Note that since the parameter ε is approximately equal to $2x_{\max}/R$, it is indeed small under the assumption we made above.

3.1.1.3 Solution by means of regular perturbation

We shall solve the equation (3.4) using regular perturbation according to the recipe above, and we start by putting

$$x(t) = x_0(t) + x_1(t)\varepsilon + x_2(t)\varepsilon^2 + \cdots$$

into the equation:

$$\ddot{x} = \ddot{x}_0 + \ddot{x}_1\varepsilon + \ddot{x}_2\varepsilon^2 + \cdots = -(1+\varepsilon x)^{-2}$$

$$= -\left[1 + (-2)\varepsilon x + \frac{(-2)(-3)}{2}(\varepsilon x)^2 + \cdots\right]$$

$$= -1 + 2\varepsilon(x_0 + \varepsilon x_1 + \ldots) - 3\varepsilon^2 x_0^2 + \cdots \tag{3.6}$$

$$= -1 + \varepsilon 2x_0 + \varepsilon^2(2x_1 - 3x_0^2) + \cdots$$

(Note the use of Newton's binomial theorem). By collecting the coefficients in front of each power of ε, we find the system

$$\ddot{x}_0 = -1, \tag{3.7}$$

$$\ddot{x}_1 = 2x_0, \tag{3.8}$$

$$\ddot{x}_2 = 2x_1 - 3x_0^2, \tag{3.9}$$

$$\ddot{x}_3 = 2x_2 + 2x_0 x_1 - 2x_0\left(2x_1 + x_0^2\right) - 2\left(2x_1 - 3x_0^2\right)x_0, \tag{3.10}$$

$$\vdots$$

To find the last equation, we had to expand the series in (3.6) to order ε^3. We must also decide what to do with the initial conditions, but here it is reasonable to use

$$x_0\left(0\right) = 0,\ \dot{x}_0\left(0\right) = 1,$$

$$x_1\left(0\right) = 0,\ \dot{x}_1\left(0\right) = 0,$$

$$x_2\left(0\right) = 0,\ \dot{x}_2\left(0\right) = 0,$$

$$\vdots$$

Thus, x_0 takes care of the initial conditions, which are consequently satisfied no matter where we stop the series expansion. The solution for x_0 follows immediately from equation (3.7):

$$x_0\left(t\right) = t - \frac{1}{2}t^2.$$

By introducing this into the next equation in (3.8), we find

$$\ddot{x}_1 = 2x_0 = 2\left(t - \frac{1}{2}t^2\right),$$

or

$$x_1\left(t\right) = \frac{1}{3}t^3 - \frac{1}{12}t^4.$$

Note that only the particular solutions change with every step, and that the contribution from the homogeneous solutions disappear for $x_i\left(t\right)$ when $i \geq 1$.

Usually, the algebra quickly becomes quite complicated, but today we can make good use of software for symbolic manipulation, such as *Maple, MuPad, Mathematica,* or the free *wxMaxima* (*Maple* was used here):

$$x(t) = t - \frac{1}{2}t^2 + \varepsilon\left(\frac{1}{3}t^3 - \frac{1}{12}t^4\right) + \varepsilon^2\left(-\frac{1}{4}t^4 + \frac{11}{60}t^5 - \frac{11}{360}t^6\right) + \mathcal{O}\left(\varepsilon^3\right).$$

(3.11)

From this solution, we also find a more accurate equation for the time to the maximum,

$$\frac{d}{dt}\left(t - \frac{1}{2}t^2 + \varepsilon\left(\frac{1}{3}t^3 - \frac{1}{12}t^4\right) + \varepsilon^2\left(-\frac{1}{4}t^4 + \frac{11}{60}t^5 - \frac{11}{360}t^6\right)\right)$$

$$= 1 - t + \varepsilon t^2 - \frac{1}{3}\varepsilon t^3 - \varepsilon^2 t^3 + \frac{11}{12}\varepsilon^2 t^4 - \frac{11}{60}\varepsilon^2 t^5 = 0. \qquad (3.12)$$

This is a fifth-degree equation, but since we expect the solution t_m to be close to 1, we can, in accordance with the foregoing, try a perturbation expansion:

$$t_m = 1 + a\varepsilon + b\varepsilon^2 + \mathcal{O}(\varepsilon^3).$$

By introducing this into (3.12), we find

$$0 = 1 - \left(1 + a\varepsilon + b\varepsilon^2\right) + \varepsilon\left(1 + a\varepsilon + b\varepsilon^2\right)^2 - \frac{1}{3}\varepsilon\left(1 + a\varepsilon + b\varepsilon^2\right)^3$$

$$- \varepsilon^2\left(1 + a\varepsilon + b\varepsilon^2\right)^3 + +\frac{11}{12}\varepsilon^2\left(1 + a\varepsilon + b\varepsilon^2\right)^4 - \frac{11}{60}\varepsilon^2\left(1 + a\varepsilon + b\varepsilon^2\right)^5 + \cdots$$

$$= \left(-a + \frac{2}{3}\right)\varepsilon + \left(a - \frac{4}{15} - b\right)\varepsilon^2 + \mathcal{O}\left(\varepsilon^3\right).$$

This gives up to $\mathcal{O}\left(\varepsilon^2\right)$

$$a = \frac{2}{3}, \quad b = a - \frac{4}{15} = \frac{2}{5},$$

and

$$t_m = 1 + \frac{2}{3}\varepsilon + \frac{2}{5}\varepsilon^2 + \mathcal{O}(\varepsilon^3). \qquad (3.13)$$

It is reasonable that the time it takes up to a maximum increases a little from 1, since gravity acting on the projectile becomes weaker as it rises. Figure 3.1 shows some numerical solutions created using *Matlab*[TM].

The perturbation expansion to zeroth, first, and second order is compared to the numerical solution in the Figures 3.2 and 3.3.

3.1.1.4   Analytical solution

As remarked above, it is not possible to express the full solution of

$$\ddot{x} = -\frac{1}{(1 + \varepsilon x)^2}, \quad x(0) = 0, \quad \dot{x}(0) = 1,$$

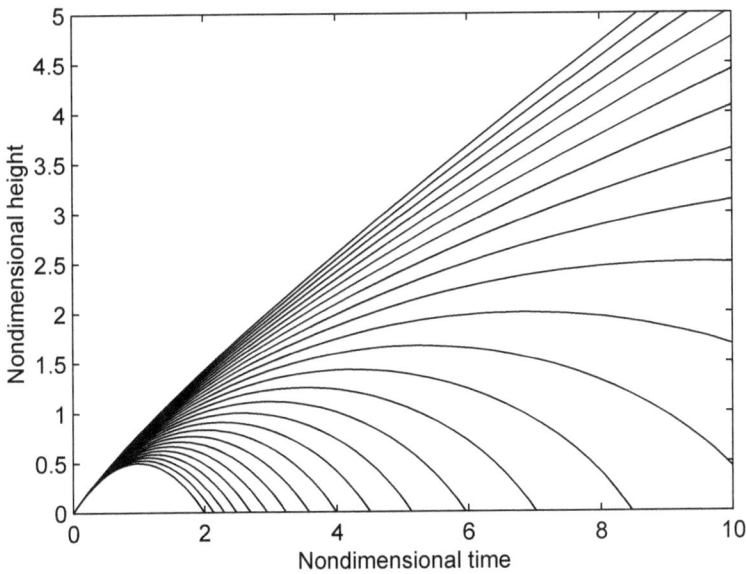

FIGURE 3.1
Numerical solutions shown for $\varepsilon \in [0, 3]$. When $\varepsilon \geq 2$, the initial speed is above the escape speed, and the projectile never returns to $x = 0$.

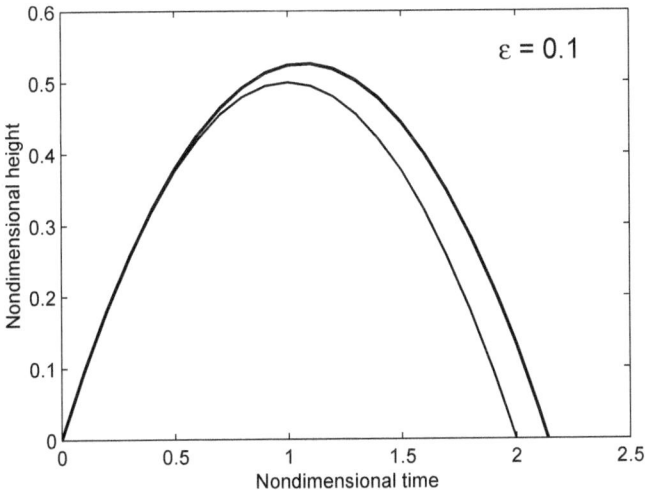

FIGURE 3.2
Numerical solution and perturbation solutions for $\varepsilon = 0.1$. The numerical solution (thick line) and the perturbation solutions to first- and second-order collapse on the graph. The solution for x_1 is different, however.

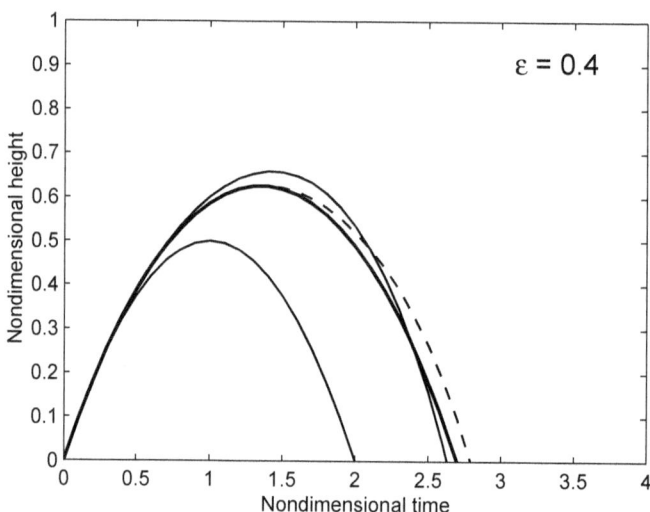

FIGURE 3.3
Similar to the previous figure for $\varepsilon = 0.4$. The thick curve represents the numerical solution, the thin curve represents x_0 and $x_0 + \varepsilon x_1$, and the dashed curve represents $x_0 + \varepsilon x_1 + \varepsilon^2 x_2$.

in closed form by means of elementary functions. However, it is possible to do something. After multiplying the equation with $\dot{x}$, we find

$$\frac{\mathrm{d}}{\mathrm{d}t}\left(\dot{x}^2/2\right) = \frac{\mathrm{d}}{\mathrm{d}t}\left(\frac{1}{\varepsilon}\frac{1}{1+\varepsilon x}\right),$$

or that

$$\frac{\dot{x}^2}{2} - \frac{1}{\varepsilon}\frac{1}{1+\varepsilon x} = \text{constant.}$$

This tells us that the motion is *conservative*. By introducing the initial conditions $x(0) = 0$ and $\dot{x}(0) = 1$, we find that the constant is equal to $1/2 - 1/\varepsilon$. This leads to a first-order non-linear equation,

$$\dot{x}^2 = \frac{1 + (\varepsilon - 2)x}{1 + \varepsilon x}, \quad x(0) = 0. \tag{3.14}$$

If $\varepsilon < 2$, $\dot{x}$ will be 0 for

$$x_{\max} = \frac{1}{2 - \varepsilon}.$$

This is therefore the exact expression for the maximum height of the projectile when $\varepsilon < 2$. If $\varepsilon > 2$, the speed will always be greater than 0, and the projectile continues to the boundaries of the universe. Note that for Earth, $\varepsilon = 2$ means that

$$V = \sqrt{2Rg} \approx \sqrt{2 \times \frac{40000000}{2\pi} \times 9.81\frac{\mathrm{m}}{\mathrm{s}}} \approx 11.17\mathrm{km/s},$$

which is therefore the escape velocity referred to above.

By taking the square root of (3.14) and separating the variables, and as long as $\dot{x} \geq 0$, we may write the solution of (3.14), implicitly as

$$t = \int_0^x \sqrt{\frac{1 + \varepsilon s}{1 + (\varepsilon - 2)s}} ds.$$

This integral turns out to be solvable,

$$\int \sqrt{\frac{a + s}{b - s}} ds = \frac{a + b}{2} \arcsin\left(\frac{2s + a - b}{a + b}\right) - \sqrt{(a + s)(b - s)} + C = F(s, a, b) + C.$$

Thus,

$$t = \int_0^x \sqrt{\frac{1 + \varepsilon s}{1 - (2 - \varepsilon)s}} ds = \sqrt{\frac{\varepsilon}{2 - \varepsilon}} \int_0^x \sqrt{\frac{1/\varepsilon + \varepsilon s}{1/(2 - \varepsilon) - s}} ds$$

$$= \sqrt{\frac{\varepsilon}{2 - \varepsilon}} \left[F(x, \frac{1}{\varepsilon}, \frac{1}{2 - \varepsilon}) - F(0, \frac{1}{\varepsilon}, \frac{1}{2 - \varepsilon}) \right].$$

Since we already know that

$$x_{\max} = \frac{1}{2 - \varepsilon},$$

when $\varepsilon < 2$, we also find an exact expression for t_m:

$$t_m = \int_{s=0}^{1/(2-\varepsilon)} \sqrt{\frac{1 + \varepsilon s}{1 - (2 - \varepsilon)s}} ds = \frac{\frac{\pi}{2} - \arcsin(1 - \varepsilon) + \sqrt{(2 - \varepsilon)\varepsilon}}{(2 - \varepsilon)^{3/2}\varepsilon^{1/2}}$$

$$= 1 + \frac{2}{3}\varepsilon + \frac{2}{5}\varepsilon^2 + \frac{8}{35}\varepsilon^3 + \mathcal{O}\left(\varepsilon^4\right).$$

The start of the power series is similar to what we found in (3.13).

3.1.2   Women World Record for 100 meters

Florence Griffith Joyner, "Flo-Jo" (1959–98) was an American track runner who is still (as of this writing) the holder of the official world record for 100 meters, 10.49s. The record was set during a quarter-final of the US qualifying heats for the Seoul Olympics in 1988. The wind gauge registered 0 m/s, while many argued that there was considerable tailwind, estimated to be about 4 m/s, and that the meter did not work properly. In the rest of the qualifying races, she ran at times around 10.7s.

A sprinter is dependent on the pushing force she/he is able to produce. This force may be written Mp^*, where M is the runner's mass and p^* a parameter with the unit of acceleration. The maximum pushing force is thus MP, where P is the maximum p^* the runner is able to produce.

In addition, there are two forces slowing the sprinter: *air resistance* and *internal friction*. The internal friction, which represents the resistance of muscles and joints, is believed to be written in the form Mu^*/τ, where u^* is the runner's speed and τ is a characteristic time constant. Measurements of different runners, including Ben Johnson and Carl Lewis, have given $P \approx 10\text{m/s}^2$ and $\tau \approx 1\text{s}$, and we shall use these values below.

Based on Newton's second law, we can now write the equation of motion for the sprinter

$$M\frac{du^*}{dt^*} = Mp^*(t^*) - M\frac{u^*}{\tau} - F_l, \qquad (3.15)$$

where F_l represents the air resistance. The expression for the air resistance may be found by dimensional analysis. It is reasonable to assume that F_l depends on the air density, ρ_{air}, the air's kinematic viscosity ν, the runner's velocity u^*, and the runner's cross-sectional area A in the direction of motion. In addition, we need a length scale L, for which we may use $\sqrt{A}$. It is left to the reader to show that from dimensional analysis we may now write

$$F_l = \frac{1}{2}\rho_{\text{air}}C_D\,(Re)\,Au^{*2},$$

where C_D is an unknown function, the so-called *drag coefficient*, which depends on *Reynolds number*, $Re = \frac{\sqrt{A}u^*}{\nu}$.

Let us now scale (3.15). Since we already know that τ is a typical time constant, we decide to use this as our time scale. We know the maximum P, and because of its unit, this is a natural scale for the acceleration. Thus, we scale the velocity by using $P\tau$ and obtain

$$\dot{u}(t) + u(t) + \varepsilon u(t)^2 = p(t), \qquad (3.16)$$

where

$$\varepsilon = \tfrac{1}{2}\rho_l C_D\left(\frac{\sqrt{A}u^*}{\nu}\right)\tau^2 P\frac{A}{M}.$$

Now C_D itself is depending on the speed, but measurements of air resistance for irregular bodies have shown that C_D is almost constant for $2 \times 10^4 < Re < 10^6$, which covers mainly what we are facing here. This value of C_D is close to 1, which we, for simplicity, shall use below. With P and τ given as above, $A \approx 0.45\text{m}^2$ and $\rho_{air} \approx 1.2\text{kg/m}$, ε is thus about 0.035 for an athlete weighing 70 to 80 kg. Since $|u| \leq 1$, we conclude that the air resistance is a relatively small term in the equation.

In order to solve (3.16) by regular perturbation, we write

$$u(t) = u_0(t) + \varepsilon u_1(t) + \mathcal{O}(\varepsilon^2)$$

and put this in (3.16). Then

$$\dot{u}_0 + \varepsilon\dot{u}_1 + \mathcal{O}(\varepsilon^2) + u_0 + \varepsilon u_1 + \mathcal{O}(\varepsilon^2) + \varepsilon(u_0 + \varepsilon u_1 + \mathcal{O}(\varepsilon^2))^2 = p.$$

We collect all parts of the same order in ε and use $u(0) = 0$ as the initial condition. This gives us a sequence of first-order equations:

$$\dot{u}_0 + u_0 = p, \ u_0(0) = 0,$$
$$\dot{u}_1 + u_1 = -u_0^2, \ u_1(0) = 0,$$
$$\dot{u}_2 + u_2 = -2u_0u_1, \ u_2(0) = 0,$$

$$\vdots$$

In order to solve the equations, we must also decide what to use for the acceleration $p(t)$. Let us, for simplicity, assume that $p^*(t) = P$, that is, $p(t) \equiv 1$. The first two terms of the perturbation expansion are thus determined by

$$\dot{u}_0 + u_0 = 1, \ u_0(0) = 0,$$
$$\dot{u}_1 + u_1 = -u_0^2, \ u_1(0) = 0,$$

and we easily find that the solution to order ε is

$$u(t) = 1 - e^{-t} + \varepsilon[-1 + 2te^{-t} + e^{-2t}] + \mathcal{O}\left(\varepsilon^2\right).$$

By plotting the graph we see that the sprinter reaches maximum velocity $(u_0(t) \approx 1)$ approximately at $t = 3$, that is, after $3\tau = 3$ seconds.

Let us now assume that we have wind blowing parallel to the running direction. This leads to a modified drag

$$F_l = \tfrac{1}{2}\rho_l C_D A(u^* - W)^2.$$

Wind speed is scaled similar to the runner's speed, so that in dimensionless variables the dimensionless wind is given by $\delta = W/(\tau P)$.

Let us determine the maximum velocity U a sprinter can hold as a function of δ to the first order in ε. The maximum speed is achieved when the acceleration is zero, i.e., given by the equation

$$U + \varepsilon(U - \delta)^2 = 1.$$

Check that the solution to first order in ε is

$$U = 1 - \varepsilon(1 - \delta)^2 + \mathcal{O}(\varepsilon^2).$$

We assume that Florence exerted maximal effort in all races and, albeit somewhat unrealistically, maintained her maximum speed throughout each race. The maximum speed without wind is $U_0 = 1 - \varepsilon$, while $U_4 = 1 - 0.36\varepsilon$ with a tailwind equal to 4m/s, that is, $\delta = 0.4$. The total time used for 100m will be $T = (100\text{m})/U$, and thus

$$T_0/T_4 = (1 - 0.36\varepsilon)/(1 - \varepsilon).$$

To get an idea of what this means in time, we must find "her" ε. We assume that she had the same acceleration, cross-sectional area and drag coefficient as the one above. Her weight, however, should be somewhat less, let's say 60 kg. That gives $\varepsilon \approx 0.04$. Expressed in time 4 m/s tailwind gives about

$$10.7\text{s} \cdot \frac{1 - 0.04}{1 - 0.04 \times 0.36} = 10.42\text{s}.$$

As she accelerated for the first three seconds, this fits very well with the time she actually used. It should also be noted that she ran at 10.54s in the finals, but then the tailwind was about 2m/s (was that reasonable?).

3.1.3   Exercises

1. Case B in the discussion of the falling sphere in a fluid led to the equation

    $$2\ddot{x} + \varepsilon\dot{x} = 1, \ x(0) = 0, \ \dot{x}(0) = 0, \ 0 < \varepsilon \ll 1.$$

    This equation has the exact solution

    $$x_{\text{sol}}(t) = \frac{2}{\varepsilon^2}\left(e^{-\varepsilon t/2} - 1\right) + \frac{t}{\varepsilon}.$$

    (a) Determine x_0, x_1, and x_2 in the regular perturbation expansion

    $$x(t) = x_0(t) + \varepsilon x_1(t) + \varepsilon^2 x_2(t) + \cdots,$$

    and show that it agrees with the start of the power series development in ε of the exact solution.

    (b) An approximate solution $x_a(t, \varepsilon)$ is a *uniform approximation* to the exact solution, x_{sol}, on the interval $[0, 1]$ if

    $$\lim_{\varepsilon \to 0}\left(\max_{t \in [0,1]} |x_a(t, \varepsilon) - x_{\text{sol}}(t)|\right) = 0.$$

    Does this apply to $x_a(x, \varepsilon) = x_0(t) + \varepsilon x_1(t)$? What if we replace $[0, 1]$ with $[0, \infty)$?

2. Consider the problem

    $$y''(t) + \varepsilon y'(t) + 1 = 0, \tag{3.17}$$
    $$y(0) = 0, \ y'(0) = 0, \ \ 0 < \varepsilon \ll 1. \tag{3.18}$$

    Determine the start of the perturbation expansion $y_0(t) + \varepsilon y_1(t) + \varepsilon^2 y_2(t)$ to the solution for $t \geq 0$. Compare to the exact solution. (*Hint*: The general solution of Eq. 3.17 has the general form $y(t) = A + Be^{-\varepsilon t} - t/\varepsilon$.)

3. This problem is somewhat similar to the sphere falling in a fluid (the scaling model problem without gravity), but in this case the friction is more realistic and nonlinear. The equation may be written

$$m\frac{dv^*}{dt^*} = -av^* + bv^{*2}, \quad v^*(0) = V_0.$$

We have been told that $a, b > 0$, and also that $bV_0 \ll a$.

(a) Find the (obvious) scale for v^* and then the scale for time, T, from the simplified equation, $m\frac{dv^*}{dt^*} = -av^*$, by the *rule of thumb*

$$T = \frac{\max|v^*(t)|}{\max|dv^*(t)/dt^*|}.$$

Show that this scaling leads to the equation

$$\frac{dv}{dt} = -v + \varepsilon v^2,$$

$$v(0) = 1, \quad \varepsilon \ll 1.$$

(b) Determine v_0 and v_1 in the series expansion $v(t) = v_0(t) + \varepsilon v_1(t) + \cdots$. Is this result reasonable for all $t > 0$ when the general solution of $\dot{y} + y - \varepsilon y^2 = 0$ is

$$y(t) = \frac{e^{-t}}{C + \varepsilon e^{-t}}?$$

4. During the modelling of the sprinters, we derived the equation

$$M\frac{du^*}{dt^*} = Mp^*(t^*) - M\frac{u^*}{\tau} - F_{\text{air}},$$

where M is the runner's mass, u^* the velocity, p^* a "performance variable", τ a time constant, and F_{air} the air resistance.

(a) Explain why the term for air resistance, found from dimensional analysis, ought to be stated as

$$F_{\text{air}} = \tfrac{1}{2}\rho_{\text{air}}C_D\,(Re)\,A(u^* - W)\,|u^* - W|.$$

(Here ρ_{air} is the air density, Re the Reynolds number and A the runner's cross-sectional area). After scaling,

$$p^* = Pp,$$

$$t^* = \tau t,$$

$$u^* = (P\tau)\,u,$$

and without wind, the equation becomes

$$\dot{u}(t) + u(t) + \varepsilon u(t)^2 = p(t), \tag{3.19}$$

$$u(0) = 0, \tag{3.20}$$

where

$$\varepsilon = \tfrac{1}{2}\rho_l C_D\,(Re)\,\tau^2 P\tfrac{A}{M}.$$

(b) Estimate ε for Usain Bolt and Florence Griffith-Joyner when we assume here and below that $\rho_{\text{air}} = 1.2\text{kg/m}^3$, $C_D\,(Re) \equiv 1$, $P = 10\text{m/s}^2$, and $\tau = 1\text{s}$.

(c) Verify that the solution of (3.19) with $p\,(t) \equiv 1$ will be

$$u(t) = 1 - e^{-t} + \varepsilon[-1 + 2te^{-t} + e^{-2t}] + \mathcal{O}\left(\varepsilon^2\right).$$

(d) We scale the wind speed in the same way as u^*, so that

$$u^* - W = (P\tau)\,(u - \delta).$$

Show that the dimensionless maximum speed $u_{\max}$ (when $u_{\max} > \delta$) is given by

$$u_{\max} + \varepsilon(u_{\max} - \delta)^2 = 1.$$

(e) Show that in order to determine the more exact time spent on running, T_a, on the basis of this model, it will be necessary to solve the equation

$$\frac{L}{P\tau^2} = \int_0^{T_a/\tau} u\,(t)\,\mathrm{d}t,$$

where $L = 100\text{m}$.

(f) The discussion in the last few paragraphs of the course note is rough, since it is assumed that the runner holds a maximum speed during the whole distance. The advantage of the tailwind is therefore estimated to be too large. From the information about her time under controlled conditions, it is possible to derive the size of ε, provided that the model holds. So what is the conclusion of this study? *Tailwind* or *doping*?

MATLAB code for numerical experiments:

```
% SCRIPT
global EPS DELTA
% Parameters:
P   = 10;    % Maximum performance factor [m/s^2]
tau = 1.0;   % Relaxation time [s]
M   = 60;    % Body mass [kg]
A   = 0.40;  % Cross sectional area [m^2]
rho = 1.2;   % Density [kg/m^3]
W   = 0 ;    % Wind speed [m/s]
Cd  = 1;     % Drag coefficient
%
EPS   = 0.5*rho*Cd*tau^2*P*A/M
DELTA = W/(P*tau)
```

```
%
Treal = 0:0.1:12; % NB! Time in seconds
tspan = Treal/tau;
% ODE solver
[t,y] = ode45(@FJfunc,tspan,[0 0]');
T = t*tau;          % Real time
L = P*tau^2*y;      % Real distance
plot(T,L); xlabel('Time [s]') ; ylabel('Distance [m]');
legend('Position','Velocity')
grid

function dydt = FJfunc(t,y)
global EPS DELTA
dydt     = [0 0]';
dydt(1) = y(2);
dydt(2) = 1-y(2)-EPS*(y(2)-DELTA)*abs(y(2)-DELTA);
```

3.2 Singular Perturbation

Regular perturbation expansions with respect to a small parameter ε do not always produce a complete approximate solution. In fact, there are several situations where this method fails, in particular:

1. When the small parameter multiplies the highest derivative in an ordinary differential equation problem.

2. When algebraic equations change degree when $\varepsilon = 0$.

3. When $\varepsilon = 0$ changes the character of the problem, as in the case of a partial differential equation changing type (from elliptic to parabolic, for example). In other words, the solution for $\varepsilon = 0$ is fundamentally different in character from the solution for ε close to zero.

4. When the problems occur on infinite domains, giving rise to so-called *secular* terms.

Such problems fall in the general category of singular perturbation, where a common feature is that the equations have multiple time or spatial scales, each covering only a part of the solution domain.

For differential equations, in particular in fluid mechanics, problems involving *boundary layers* are common. If there is a boundary layer, a leading-order regular perturbation term found by setting $\varepsilon = 0$ in the equations often provides a valid approximation in a large *outer* region but fails severely in the boundary layer. A valid approximation in the boundary layer is found by a *rescaling* of the equations, as discussed below. Often the inner and outer approximations with respect to the boundary layer may be *matched* to obtain a uniformly valid approximation over the entire domain of interest. The singular perturbation method applied in this context is also called the method of *matched asymptotic expansions*.

The rest of this section provides a simple introduction to singular perturbation methods, but the reader should be aware that it is a huge subject of which we only cover a tiny fraction. Parts of the analysis and the example equations have been adopted from [31].

3.2.1   Algebraic equations

Consider the quadratic equation

$$\varepsilon x^2 + 2x + 1 = 0, \ \ 0 < \varepsilon \ll 1.$$

The equation can easily be solved exactly, but our goal is to illustrate the failure of the regular perturbation method in this particular case. When $x = \mathcal{O}(1)$, the equation is a small modification of the unperturbed equation

$$2x + 1 = 0,$$

which is linear and has the single solution $x = -1/2$. It is thus fundamentally different from the full quadratic equation. If we now attempt to use the regular perturbation method by substituting the series

$$x = x_0 + \varepsilon x_1 + \varepsilon^2 x_2 + \cdots.$$

into the equation, we obtain, after setting the coefficients of each power of ε equal to zero, a set of equations

$$2x_0 = -1,$$
$$2x_1 = -x_0^2,$$
$$2x_2 = -2x_1 x_0,$$
$$\vdots$$

Solving the equations in sequence leads to

$$x = -\frac{1}{2} - \frac{1}{8}\varepsilon - \frac{1}{16}\varepsilon^2 + \cdots$$

(Try to derive this yourself). What happens to the other solution? Regular perturbation assumes the leading term is of order unity, and it is not surprising that here it recovers only one root of order unity. To find the second root, we examine the three terms, εx^2, $2x$, and 1, of the equation more closely. Discarding the εx^2-term gave a root close to $x = -1/2$, and in that case, the term εx^2 is indeed small compared to $2x$ and 1. Thus, it is reasonable to ignore εx^2 and the apparently small term is indeed small.

Clearly, we have to search for the other solution for x-es where εx^2 may not be small compared to the other terms. Our first guess could be to assume $x = \mathcal{O}\left(\varepsilon^{-1/2}\right)$, say $x = a/\varepsilon^{1/2}$ where $a = O(1)$. This leads to

$$\varepsilon x^2 = a^2 = \mathcal{O}(1),$$
$$2x = a/\varepsilon^{1/2} = \mathcal{O}\left(\varepsilon^{-1/2}\right),$$
$$1 = \mathcal{O}(1).$$

The second term dominates, and this scaling leads nowhere.

Let us then try $x = a/\varepsilon$, again assuming $a = \mathcal{O}(1)$. Then

$$\varepsilon x^2 = a^2/\varepsilon = \mathcal{O}\left(\varepsilon^{-1}\right),$$
$$2x = a/\varepsilon = \mathcal{O}\left(\varepsilon^{-1}\right),$$
$$1 = \mathcal{O}(1).$$

The first two terms dominate, and multiplying through by ε, we obtain

$$\varepsilon\left(\varepsilon x^2 + 2x + 1\right) = a^2 + 2a + \varepsilon = 0.$$

This makes sense, and gives us a quadratic equation for a amenable to regular perturbation. Check, using regular perturbation, that

$$a^2 + 2a + \varepsilon = 0$$

has solutions

$$a_I = 0 - \frac{1}{2}\varepsilon + \mathcal{O}\left(\varepsilon^2\right),$$
$$a_{II} = -2 + \frac{1}{2}\varepsilon + \mathcal{O}\left(\varepsilon^2\right),$$

leading to

$$x_I = \frac{1}{\varepsilon}\left(-\frac{1}{2}\varepsilon + \mathcal{O}\left(\varepsilon^2\right)\right) = -\frac{1}{2} + \mathcal{O}(\varepsilon),$$
$$x_{II} = \frac{1}{\varepsilon}\left(-2 + \frac{1}{2}\varepsilon + \mathcal{O}\left(\varepsilon^2\right)\right) = -\frac{2}{\varepsilon} + \frac{1}{2} + \mathcal{O}(\varepsilon).$$

In this case, x_I is just the leading order solution we found by straightforward regular perturbation, whereas the second one required further investigations. The crucial assumption turned out to be to assume $x = \mathcal{O}\left(\varepsilon^{-1}\right)$, and this is actually a *rescaling* of x. The rescaling leads to changes in the relative size of the terms in the equation, and a successful rescaling gives us equations where we may again apply regular perturbation.

In the present case, it is possible to expand the exact solutions and show that the approximate solutions are reasonable by recalling that $\sqrt{1-\varepsilon} = 1 - \frac{1}{2}\varepsilon + O\left(\varepsilon^2\right)$:

$$
\begin{aligned}
x_{I,II} &= \frac{1}{2\varepsilon}\left(-2 \pm \sqrt{4 - 4\varepsilon}\right) \\
&= \frac{1}{\varepsilon}\left(-1 \pm \sqrt{1 - \varepsilon}\right) \\
&= \frac{1}{\varepsilon}\left(-1 \pm \left(1 - \frac{1}{2}\varepsilon + O\left(\varepsilon^2\right)\right)\right) \\
&= \begin{cases} -\frac{1}{2} + \mathcal{O}\left(\varepsilon\right), \\ -\frac{2}{\varepsilon} + \frac{1}{2} + \mathcal{O}\left(\varepsilon\right). \end{cases}
\end{aligned}
$$

In summary, roots may be of different orders and one expansion does not reveal them all. The reasoning we used in this example is called *dominant balancing*. We examine each term carefully and determine which ones combine to give a dominant balance.

As another example, let us try to find the leading order approximation of the four roots to the equation

$$\varepsilon x^4 - x - 1 = 0, \ 0 < \varepsilon \ll 1.$$

For $\varepsilon = 0$ we obtain only the single root $x = -1$ of order one. To determine the leading order of the other roots, we again use dominant balancing.

The first and third terms balance if $x = \mathcal{O}(\varepsilon^{-1/4})$, which, however, makes the second term $\mathcal{O}(\varepsilon^{-1/4})$ dominate. The first and second terms balance for $x = O(\varepsilon^{-1/3})$ and now both these terms are large compared to 1. Hence, we rescale according to

$$x = a\varepsilon^{-1/3},$$

which gives

$$a^4 - a - \varepsilon^{1/3} = 0. \tag{3.21}$$

To the leading order, $a^4 - a = 0$, with solutions $a = 0, 1, \ e^{2\pi i/3}, \ e^{-4\pi i/3}$.

We discard $a = 0$ because that corresponds to the $\mathcal{O}(1)$-solution $x = -1$. Consequently, to leading order the four roots are

$$x = -1, \ \varepsilon^{-1/3}, \ \varepsilon^{-1/3}e^{2\pi i/3}, \ \varepsilon^{-1/3}e^{-4\pi i/3}.$$

Higher-order approximations to the $\mathcal{O}(1)$-solution obtained by a regular perturbation series in powers of ε, whereas for (3.21) we need a power series in $\varepsilon^{1/3}$.

3.2.2   Scalar differential equations

Consider the following boundary value problem on $0 \le x \le 1$ when $0 < \varepsilon \ll 1$:

$$\varepsilon y''(x) + (1 + \varepsilon)y'(x) + y(x) = 0, \tag{3.22}$$
$$y(0) = 0, \ y(1) = 1.$$

By observing that the characteristic equation, $\varepsilon r^2 + (1 + \varepsilon)r + 1 = 0$, has solutions $r_{1,2} = -1, -1/\varepsilon$, the problem is easily solved analytically, and we leave it to the reader to show that the exact solution is

$$y_s(x) = \frac{e^{-x} - e^{-x/\varepsilon}}{e^{-1} - e^{-1/\varepsilon}}. \tag{3.23}$$

However, if we assume a regular perturbation solution

$$y = y_0(x) + \varepsilon y_1(x) + \varepsilon^2 y_2(x) + \ldots,$$

substitution into (3.22) leads to the system

$$y_0' + y_0 = 0, \tag{3.24}$$
$$y_1' + y_1 = -y_0'' - y_0', \tag{3.25}$$

$$\vdots \tag{3.26}$$

Since the general solution of (3.24) is just $y_0(x) = Ce^{-x}$, there is no way y_0 can satisfy both boundary conditions. If we require that $y_0(0) = 0$, the solution of (3.24) is simply $y_0(x) = 0$, and if we require $y_0(1) = 1$, the solution is $y_0(x) = e^{1-x}$. Thus, contrary to the regular perturbation case, it is impossible for the leading-order perturbation solution to satisfy both boundary conditions. Since there are two boundary conditions, two linearly independent solutions of (3.22) are required, and (3.24) can only provide one such solution, which, in this particular case, is also a solution of (3.22). There is no obvious way to get the second solution by regular perturbation. In particular, a rescaling of y does not help us at all. It turns out that what is needed is a rescaling of x.

The exact solution for a set of ε values is shown in Figure 3.4. The solution is seen to vary smoothly apart from an interval near $x = 0$. In fact, as ε gets smaller, the term $\varepsilon y''(x)$, which is supposed to be small, is not small at all. It is instructive to estimate the size for each term in the equation for, say, $x = \varepsilon$:

$$|y_s(\varepsilon)| = \mathcal{O}(1),$$
$$|(1 + \varepsilon)y_s'(\varepsilon)| = \mathcal{O}(\varepsilon^{-1}),$$
$$|\varepsilon y_s''(\varepsilon)| = \mathcal{O}(\varepsilon^{-1}).$$

(Check it!) For $x = \mathcal{O}(\varepsilon)$, the first and second derivative terms are dominating the equation, whereas for $x = 1/2$ say, all derivatives are $\mathcal{O}(1)$ and the second derivative term is $\mathcal{O}(\varepsilon)$.

The region near to the origin is called the *boundary layer* or the *inner region*, and the rest of the interval, away from the origin, is called *outer region*. Only one scaling does not cover the behaviour in both layers. Rather, two scalings are needed, one for each region.

Thus, we see again that terms appearing to be small and insignificant are not necessarily so. However, for values of x away from the boundary layer, the

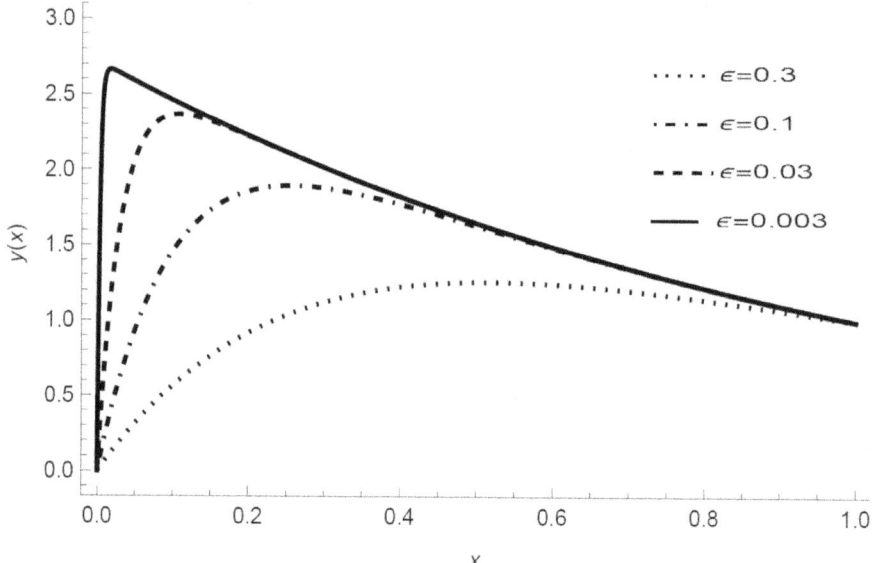

FIGURE 3.4
Exact solutions showing a narrow layer near $x = 0$ where rapid changes occur
when ε becomes small.

term $\varepsilon y''$ is indeed small. This suggests that in the outer region, the leading-
order equation (3.24), obtained by setting $\varepsilon = 0$ in the original problem, is
a valid approximation, provided we only take the right $(x = 1)$ boundary
condition into account. Thus, $y_0(x)$ should be e^{1-x}. This is indeed consistent
with the exact solution in Eq. 3.23 as long as $x = \mathcal{O}(1)$, since for small ε,
$e^{-1} - e^{\frac{-1}{\varepsilon}} \approx e^{-1}$ and y_s may be approximated by

$$y_s(x) = \frac{1}{e^{-1} - e^{-1/\varepsilon}} \left(e^{-x} - e^{-x/\varepsilon} \right) \approx \frac{1}{e^{-1}} e^x = e^{1-x}.$$

What about an approximate solution in the boundary layer near $x = 0$? Again
the exact solution gives us a hint: When x is $\mathcal{O}(\varepsilon)$,

$$y_s(x) \approx \frac{1}{e^{-1}} \left(1 - e^{-x/\varepsilon} \right) = e - e^{1-x/\varepsilon}.$$

Let us see what may come out of a rescaling of x for this case, and we try
$x = \varepsilon \xi$. In order to distinguish from $y(x)$, we write $Y(\xi)$ in the equation:

$$\varepsilon y''(x) + (1 + \varepsilon)y'(x) + y(x)$$

$$= \varepsilon \frac{1}{\varepsilon^2} \frac{d^2 Y}{d\xi^2} + (1 + \varepsilon)\frac{1}{\varepsilon}\frac{dY}{d\xi} + Y = 0.$$

After multiplication with ε, we thus obtain the rescaled equation

$$\frac{d^2Y}{d\xi^2} + (1+\varepsilon)\frac{dY}{d\xi} + \varepsilon Y = 0,$$

with the leading order equation

$$\frac{d^2Y_0}{d\xi^2} + \frac{dY_0}{d\xi} = 0. \tag{3.27}$$

The solution passing through 0 is clearly

$$Y_0 = A\left(1 - e^{-\xi}\right).$$

However, fitting A by requiring $Y_0\,(\xi = 1/\varepsilon) = 1$, and hence $A \approx 1$, does not at all approximate the exact solution! In fact, the boundary layer solution can *not* be used outside the boundary layer. We have to conclude that rescaling is not sufficient to determine what we call the leading order *inner solution* completely. If we return to the exact solution, $y'_s\,(0) \approx \frac{e}{\varepsilon}$, suggesting that $A \approx e$.

In this problem we have the advantage of knowing the exact solution, and as yet we have little clue how to determine the best inner solution if the exact solution is unknown.

3.2.2.1 Boundary layer analysis and matching

We return to the boundary value problem in (3.22). By examining the exact solution in a boundary layer near $x = 0$, we found that rapid changes were occurring in y_s and its derivatives. In particular, the term $\varepsilon y''$ was not small, as it appears to be when we just look at the equation. However, away from the boundary layer, in the region where $x = \mathcal{O}(1)$, it was noted that $\varepsilon y''$ and $\varepsilon y'$ are indeed small, and so the solution could be approximated accurately by setting $\varepsilon = 0$ in the equation to obtain

$$y' + y = 0.$$

and selecting the boundary condition $y(1) = 1$. This gives the outer approximation

$$y_0(x) = e^{1-x}.$$

To analyze the behaviour in the boundary layer, we notice that significant changes in y take place on a very short spatial interval, suggesting a length scale depending on ε, say $\delta\,(\varepsilon)$. Thus, if we change the independent variable x to $\xi = \frac{x}{\delta(\varepsilon)}$, and introduce $Y\,(\xi) = y\,(\delta\,(\varepsilon)\,\xi)$, the differential equation becomes

$$\frac{\varepsilon}{\delta\,(\varepsilon)^2}Y''\,(\xi) + \frac{1}{\delta\,(\varepsilon)}Y'\,(\xi) + \frac{\varepsilon}{\delta\,(\varepsilon)}Y'\,(\xi) + Y(\xi) = 0, \tag{3.28}$$

where the prime denotes derivatives with respect to ξ. Another way of looking at the rescaling is to regard $\xi = x/\delta\,(\varepsilon)$ as a scale transformation that permits examination of the boundary layer close up, as under a microscope. Hence, $\varepsilon, \delta(\varepsilon) \ll 1$.

The coefficients of the four terms in the differential equation are

$$\frac{\varepsilon}{\delta(\varepsilon)^2}, \frac{1}{\delta(\varepsilon)}, \frac{\varepsilon}{\delta(\varepsilon)}, 1.$$

In order to determine the correct scale factor $\delta(\varepsilon)$ to use, we estimate their magnitude by considering all possible dominant balancing terms in (3.28). We always include the first term because it was deleted in the outer layer approximation, and it is known that it plays a significant role in the boundary layer. In the present case, it turns out to be impossible to obtain dominant balancing of three terms (if all four terms should turn out to be equally important, no simplification can be made at all). Therefore, there are three cases to consider ("$\sim$" denotes "of the same order of magnitude as"):

$$(i) \quad \frac{\varepsilon}{\delta(\varepsilon)^2} \sim \frac{1}{\delta(\varepsilon)},$$
$$(ii) \quad \frac{\varepsilon}{\delta(\varepsilon)^2} \sim \frac{\varepsilon}{\delta(\varepsilon)},$$
$$(iii) \quad \frac{\varepsilon}{\delta(\varepsilon)^2} \sim 1.$$

Now, (ii) implies that $\delta(\varepsilon) = \mathcal{O}(1)$, thus contradicting the assumption. Moreover, for (iii), $\delta(\varepsilon) \sim \sqrt{\varepsilon}$ implies that the $1/\delta(\varepsilon)$-term dominates the whole equation. Only case (i) is in fact feasible, implying that we may take $\delta(\varepsilon) = \varepsilon$.

Thus, the differential equation scaled for the boundary layer becomes

$$Y'' + Y' + \varepsilon Y' + \varepsilon Y = 0. \tag{3.29}$$

The equation is now amenable to regular perturbation with a leading order solution as found above, $Y_0(\xi) = A + Be^{-\xi}$. Because the boundary layer is located near $x = 0$, it is reasonable to apply the boundary condition $y(0) = Y_0(0) = 0$, leading to $Y_0(\xi) = A(1 - e^{-\xi})$

In summary, we have obtained the outer approximate solution $y_0(x) = e^{1-x}$ for $x = \mathcal{O}(1)$ and the inner or boundary layer approximation $Y_0(x) = A\left(1 - e^{-x/\varepsilon}\right)$ for $x = \mathcal{O}(\varepsilon)$, each valid for an appropriate range of x. It remains to determine the constant A, which is going to be accomplished by the process of matching.

Matching

In singular perturbation theory, *matching* means to patch or glue approximate solutions together so as to obtain continuous approximate solutions that are uniformly valid on the union of the domains for the approximate solutions included in the matching.

For the present example, the matching consists of gluing together the inner and outer approximate solutions in order to obtain approximations valid on the entire interval, $0 \le x \le 1$.

It is reasonable that the inner and outer approximations should agree to some order in a domain intermediate to the boundary layer and the outer region. If $x = \mathcal{O}(\varepsilon)$, then x is in the boundary layer, and if $x = \mathcal{O}(1)$, then x is in the outer region. Thus, the overlap domain could be characterized as values of x, e.g., for which $x = \mathcal{O}(\sqrt{\varepsilon})$, since $\sqrt{\varepsilon}$ is between ε and 1, and $\varepsilon \ll \sqrt{\varepsilon} \ll 1$. In simple terms, we consider x-s that are far out as seen from the boundary layer, and far into the boundary layer as seen from the outer region. The intermediate scale suggests a new scaled independent variable η in the overlap domain, defined by $x = \eta\sqrt{\varepsilon}$. It turns out that the matching condition takes the simple form

$$\lim_{\varepsilon \to 0^+} y_0\left(\eta\sqrt{\varepsilon}\right) = \lim_{\varepsilon \to 0^+} Y_0\left(\frac{\eta\sqrt{\varepsilon}}{\varepsilon}\right)$$

Since $\eta = \mathcal{O}(1)$, we observe that the left limit is

$$\lim_{\varepsilon \to 0^+} y_0\left(\eta\sqrt{\varepsilon}\right) = \lim_{x \to 0^+} y_0(x) = y_0(0),$$

and the right hand side is

$$\lim_{\varepsilon \to 0^+} Y_0\left(\frac{\eta\sqrt{\varepsilon}}{\varepsilon}\right) = \lim_{\xi \to \infty} Y_0(\xi)$$

(assuming the limits exist). The simple rule is thus

$$\lim_{x \to 0} y_0(x) = \lim_{\xi \to \infty} Y_0(\xi). \tag{3.30}$$

For the problem defined in (3.22) the outer solution is $y_0(x) = e^{1-x}$, and $y_0(0) = e$. The inner solution is $Y_0(\xi) = A\left(1 - e^{-\xi}\right)$, and

$$\lim_{\xi \to \infty} Y_0(\xi) = A.$$

The rule now gives us the appropriate value for A, namely $A = e$.

This states that the outer approximation, as the outer variable moves into the inner region, should equal the inner approximation, as the inner variable moves into the outer region. In the following we simply apply (3.30) instead of introducing an intermediate variable.

The Uniform Approximation

In order to obtain a composite expansion uniformly valid throughout $[0, 1]$, we note that values of y_0 and Y_0 in the intermediate domain are about e. By summing the outer and inner approximations and subtracting their common limit value, we obtain

$$y^{(u)}(x) = y_0(x) + Y_0\left(\frac{x}{\varepsilon}\right) - \lim_{\xi \to \infty} Y_0(\xi)$$

$$= y_0(x) + Y_0\left(\frac{x}{\varepsilon}\right) - \lim_{x \to 0^+} y_0(x).$$

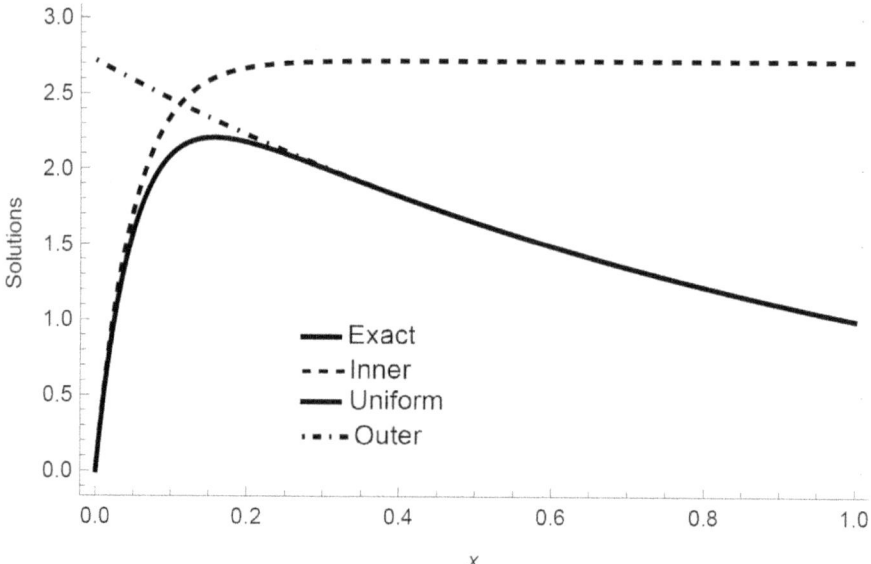

FIGURE 3.5
Exact, inner and outer solutions, and the uniform solution to the model problem in the text ($\varepsilon = 0.05$).

Convince yourself, by applying that the limits above are equal, that $y^{(u)}$ has the following properties:

$$y^{(u)}(0) = Y_0(0),$$
$$y^{(u)}(1) \approx y_0(1),$$
$$y^{(u)}(x) \approx Y_0(\frac{x}{\varepsilon}) \text{ when } x \sim \varepsilon$$
$$y^{(u)}(x) \approx y_0(x) \text{ when } x \sim 1.$$

Returning again to the problem defined in (3.22),

$$y^{(u)}(x) = y_0(x) + Y_0(x) - \lim_{\xi \to \infty} Y_0(\xi)$$
$$= e^{1-x} + e\left(1 - e^{-x/\varepsilon}\right) - e$$
$$= e^{1-x} - e^{1-\frac{x}{\varepsilon}}.$$

Note that $y^{(u)}(1)$ is not exactly equal to 1, but $e^{1-\frac{1}{\varepsilon}}$ is *very* close to 0 since $1 - \frac{1}{\varepsilon}$ is a large negative number.

Figure 3.5 shows the exact solution as well as the inner and outer solutions and the uniform solution obtained by combining the two. In this case, the uniform and the exact solutions are indistinguishable. Note that we have

only introduced approximations to leading order. Higher-order approximations can be obtained by more elaborate matching schemes, and we refer to
the references for discussions of these methods, e.g., [7].

Let us consider another simple example:

$$\varepsilon y'' + y' = 2x,$$
$$0 < x < 1, \ 0 < \varepsilon \ll 1, \qquad (3.31)$$
$$y(0) = 1, \quad y(1) = 1.$$

Clearly, regular perturbation fails since the unperturbed problem is

$$y_0' = 2x,$$

with the general solution

$$y_0(x) = x^2 + C.$$

This function cannot satisfy both boundary conditions simultaneously.

We shall assume that the boundary layer is at $x = 0$. If this assumption
fails, we could also consider a boundary layer near $x = 1$ by introducing $1 - x$
as a new independent variable. Since $x = 1$ is in the outer layer, we impose
the boundary condition $y(1) = 1$ to get the outer approximation for $x = \mathcal{O}(1)$:

$$y_0(x) = x^2.$$

To determine the rescaling in the boundary layer of width $\delta\,(\varepsilon)$, we introduce
the inner variable $\xi = \frac{x}{\delta(\varepsilon)}$ into the equation:

$$\frac{\varepsilon}{\delta\left(\varepsilon\right)^2} Y''\left(\xi\right) + \frac{1}{\delta\left(\varepsilon\right)} Y'\left(\xi\right) = 2\delta(\varepsilon)\xi.$$

If $\frac{\varepsilon}{\delta(\varepsilon)^2} \sim 2\delta(\varepsilon)$ is the dominant balance, then $\delta(\varepsilon) \sim \varepsilon^{\frac{1}{3}}$ and the second term
would be $\mathcal{O}(\varepsilon^{-\frac{1}{3}})$, which is *not* small compared to the assumed dominant
terms. We therefore assume that $\frac{\varepsilon}{\delta(\varepsilon)^2} \sim \frac{1}{\delta(\varepsilon)}$ is the dominant balance. In that
case, $\delta(\varepsilon) = \mathcal{O}(\varepsilon)$ and the term $2\delta(\varepsilon)$ is of order $\mathcal{O}(\varepsilon)$, which is indeed small
compared to $\frac{\varepsilon}{\delta(\varepsilon)^2}$ and $\frac{1}{\delta(\varepsilon)}$, since both are $\mathcal{O}(\varepsilon^{-1})$. Therefore, $\delta(\varepsilon) = \varepsilon$ is a
consistent choice and the rescaled differential equation takes the form

$$Y''\left(\xi\right) + Y'\left(\xi\right) = \varepsilon^2 \xi.$$

The inner approximation to first order satisfies

$$Y_0'' + Y_0' = 0,$$

whose general solution is

$$Y_0(\xi) = C_1 + C_2 e^{-\xi}.$$

Applying the boundary condition $y(0) = 1$ in the boundary layer gives $C_1 = 1 - C_2$, and therefore the inner approximation (with one free constant) is

$$Y_0(\xi) = (1 - C_2) + C_2 e^{-\xi}.$$

Applying the matching condition in (3.30), we find

$$\lim_{x \to 0} y_0\,(x) = 0 = \lim_{\xi \to \infty} Y_0\,(\xi) = 1 - C_2.$$

Thus, $C_2 = 1$, and $Y_0\,(\xi) = e^{-\xi}$. A uniform composite approximation $y^{(u)}\,(x)$ is found by adding the inner and outer approximations and subtracting the common limit in the overlap domain, which is zero in this case. Consequently,

$$y^{(u)}\,(x) = x^2 + e^{-\frac{x}{\varepsilon}}.$$

As an exercise, the reader should plot this approximation for different values of ε and also solve the equation analytically.

3.2.3   System of differential equations

So far, we have discussed singular perturbation for algebraic and scalar ordinary differential equations. However, systems of differential equations may also show singular perturbation behaviour, and we shall discuss some aspects of the two-dimensional theory. Consider a system

$$\dot{x} = P\,(x, y)\,, \tag{3.32}$$
$$\dot{y} = Q(x, y), \tag{3.33}$$

where the unknowns $x\,(t)$ and $y\,(t)$ are functions of t and the dot means derivative with respect to t. Since t does not enter P and Q explicitly, the system is *autonomous*. Solutions may be plotted as $x\,(t)$ and $y\,(t)$ as functions of t, but also as curves (*orbits*) parametrized by t in the (x, y)-plane, in this context called the *phase plane*. It is useful to consider the system as defining a *flow* of particles in the phase plane moving according to the velocity field $\mathbf{v} = P(x, y)\mathbf{i}_x + Q(x, y)\mathbf{i}_y$. For autonomous systems, orbits do not change with time and may be found by solving

$$\frac{dy}{dx} = \frac{Q(x, y)}{P(x, y)}.$$

Let us now consider a situation where the time scales for $x\,(t^*)$, T_x, and $y\,(t^*)$, T_y, are highly different, say $T_y \ll T_x$. This suggests that t^* should be scaled as $t^* = T_x t$ for the x-equation and $t^* = T_y \tau$ for the y-equation. After scaling, the equations may be written

$$\frac{dx}{dt} = f\,(x, y)\,,$$
$$\frac{dy}{d\tau} = g\,(x, y)\,.$$

However, when solving the system, we need to decide which of the time scales to apply. Applying $t^* = T_x t$,

$$\frac{dy}{d\tau} = \frac{dy}{dt}\frac{dt}{d\tau} = \frac{dy}{dt}\varepsilon, \quad \varepsilon = T_y/T_x \ll 1,$$

and the system becomes

$$\frac{dx}{dt} = f(x, y),$$
$$\varepsilon\frac{dy}{dt} = g(x, y). \tag{3.34}$$

Similarly, with $t^* = T_y \tau$, we obtain

$$\frac{dx}{d\tau} = \varepsilon f(x, y),$$
$$\frac{dy}{d\tau} = g(x, y). \tag{3.35}$$

Actually, these two forms correspond to the outer and inner equations above. Note that the outer system transforms to the inner system simply by introducing $\tau = t/\varepsilon$.

Consider now

$$\dot{x} = f(x, y),$$
$$\varepsilon\dot{y} = g(x, y),$$

where ε is a small positive parameter. It is customary to multiply with ε on the left-hand side, although we could alternatively have written $\dot{y} = g(x, y)/\varepsilon$. We observe that $\mathbf{v} = f(x, y)\mathbf{i}_x + \frac{g(x,y)}{\varepsilon}\mathbf{i}_y$, and unless $g(x, y) \leq \mathcal{O}(\varepsilon)$, the flow is nearly vertical with a speed proportional to ε^{-1}. This is illustrated for the linear system

$$\dot{x} = x + y,$$
$$\varepsilon\dot{y} = 1 + x - y, \quad \varepsilon = 0.01,$$

in Figure 3.6. Note that the orbits approach a path which is slightly to the side of the outer solution of the y-equation.

Let us return to solving (3.34) to leading order with initial values (x_I, y_I). Assume it is possible to solve for y from $g(x, y) = 0$, that is, $y = h(x)$, and that this solution is unique for each x. We are then left with solving

$$\dot{x}_0 = f(x_0, h(x_0)),$$

for the initial condition $x_0(0) = x_I$. The initial condition for y_0 cannot in general be satisfied since y_I will be different from $h(x_I)$.

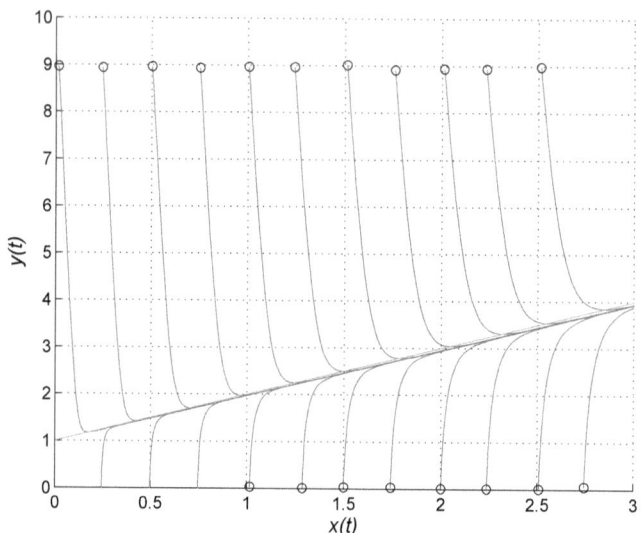

FIGURE 3.6
Flow of the simple linear system defined in the text. The initial values are indicated by small circles and the orbits quickly approach the neighborhood of the outer solution of the y-equation $(y = 1 + x)$.

Consider the inner system in (3.35) and set $X(\tau) = X_0(\tau) + \varepsilon X_1(\tau) + \cdots$, and similarly for $Y(\tau)$. To leading order,

$$\frac{\mathrm{d}X_0}{\mathrm{d}\tau} = 0,$$

$$\frac{\mathrm{d}Y_0}{\mathrm{d}\tau} = g(X_0, Y_0),$$

and thus, $X_0(\tau) = x_I$ and Y_0 are the solution of

$$\frac{\mathrm{d}Y_0}{\mathrm{d}\tau} = g(x_I, Y_0), \quad Y_0(0) = y_I.$$

For the matching of the inner and outer solutions, we first observe that $\lim_{\tau \to \infty} X_0(\tau) = x_I$ and $\lim_{t \to 0} x_0(t) = x_I$. Hence,

$$x_0^{(u)}(t) = x_0(t) + X_0\left(\frac{t}{\varepsilon}\right) - \lim_{t \to 0} x_0(t) = x_0(t) + x_I - x_I = x_0(t).$$

If $Y_0(\tau)$ tends to a limit when $\tau \to \infty$, it will usually be the case that also $\frac{\mathrm{d}Y_0}{\mathrm{d}\tau} = g(x_I, Y_0)$ tends to 0 and hence $Y_0(\tau) \to h(x_I)$, which is just what is needed for the matching. Hence,

$$y_0^{(u)}(t) = y_0(t) + Y_0\left(\frac{t}{\varepsilon}\right) - \lim_{t \to 0} y_0(t) = h(x_0(t)) + Y_0\left(\frac{t}{\varepsilon}\right) - h(x_I).$$

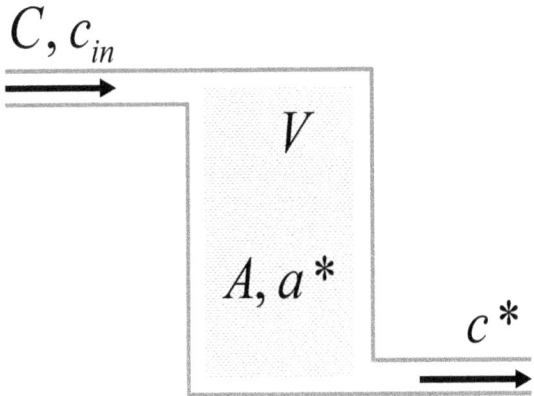

FIGURE 3.7
Sketch of a simple reaction with material A staying in the reaction tank and the solution containing C flowing through.

In the following section, we shall first apply this technique to a model of a simple physical reactor and then consider a famous dynamical system where some modifications turn out to be necessary.

3.2.4 Application I: A simple chemical reactor

A chemical reactor consists of a tank partly filled with a solid material A. A fluid containing another dissolved chemical (C) with concentration c_{in} flows into the tank. The fluid flow q is measured in volume per second and the available volume for the fluid inside the reactor is V.

The chemical reacts with the material in the tank to create a product which leaves the tank with the fluid stream. The concentration of non-converted chemical in the tank is $c^*(t^*)$, and this is also the concentration in the fluid leaving the tank. We assume that no volume changes are involved in these reactions, and the reactor is well-mixed with no significant concentration variations inside the tank.

The amount of chemical converted per time and volume unit is $k_c a^* c^*$, where a^* is the amount of available A per volume unit. However, A dissolves over time and the change in a^* per time unit is $-k_d c^* a^*$. At the start of the reaction, at $t^* = 0$, we assume that $a^*(0) = a_I$ and $c^*(0) = c_{in}$.

It follows directly from Figure 3.7 that the equation for the chemical may be expressed as

$$\frac{\mathrm{d}(Vc^*)}{\mathrm{d}t^*} = -Vk_c a^* c^* + qc_{in} - qc^*.$$

Note that all terms express the change in mass per time unit. For the dissolving material we have, similarly

$$\frac{da^*}{dt^*} = -k_d c^* a^*.$$

There are three time scales involved in this model formulation. The first-time scale T_1 is derived from the decay of the chemical in a closed volume at the beginning of the reaction where $a^* \approx a_I$,

$$\frac{dc^*}{dt^*} \approx -k_c a_I c^*.$$

Thus,

$$c^* \approx C \exp\left(-t^* / (1/k_c a_I)\right).$$

with the time scale $T_1 = 1/k_c a_I$.

The second time scale T_2 relates to the initial decay in the amount of A (when $c^* \approx c_{in}$),

$$\frac{da^*}{dt^*} = -k_d c_{in} a^*.$$

The most elegant derivation of these time scales is to follow Lin & Segel's recipe:

$$T_1 = \frac{\max |c^*|}{\max |dc^*/dt^*|} = \frac{|c_{in}|}{|k_c c_{in} a_I|} = \frac{1}{k_c a_I},$$

$$T_2 = \frac{\max |a^*|}{\max |da^*/dt^*|} = \frac{|a_I|}{|k_d c_{in} a_I|} = \frac{1}{k_d c_{in}}.$$

There is also a third time scale, $T_3 = V/q$ which is simply the time it takes to fill a volume V with the flow q. In the rest of this discussion, we shall consider a situation where $T_1 \ll T_2$.

The obvious scales for c^* and a^* are c_{in} and a_I, respectively.

When we now scale the equation by inserting $c^* = c_{in}c$, $a^* = a_I a$, and $t^* = T_1 \tau$, this leads to

$$\frac{1}{k_c a_I T_1} \frac{dc}{d\tau} = -ac + \frac{q}{k_c a_I V} (1 - c),$$

$$\frac{1}{T_1 k_d c_{in}} \frac{da}{d\tau} = -ca,$$

and the scaled equations become

$$\frac{dc}{d\tau} = -ca + \alpha (1 - c),$$

$$\frac{da}{d\tau} = -\varepsilon ca, \tag{3.36}$$

$$c(0) = 1, \ a(0) = 1, \ \tau \geq 0.$$

where

$$\varepsilon = \frac{1/k_c a_I}{1/(k_d c_{in})} = \frac{T_1}{T_2},$$

$$\alpha = \frac{T_1}{T_3}.$$

Since $T_1 \ll T_2$, ε is a small dimensionless parameter, and the system (3.36) represents a *regular* perturbation problem. Let us therefore consider the perturbation expansions

$$c(\tau) = C_0(\tau) + \varepsilon C_1(\tau) + \cdots$$
$$a(\tau) = A_0(\tau) + \varepsilon A_1(\tau) + \cdots$$

Inserting the power series in ε, we obtain the leading order

$$\frac{dC_0}{d\tau} = -C_0 A_0 + \alpha(1 - C_0),$$
$$\frac{dA_0}{d\tau} = 0. \tag{3.37}$$

In addition, the initial conditions, which may be satisfied exactly, are

$$C_0(0) = 1, \quad A_0(0) = 1.$$

Since we obviously have $A_0(\tau) = 1$, the equation for C_0 simplifies to

$$\frac{dC_0}{d\tau} + (1 + \alpha) C_0 = \alpha,$$

that is,

$$C_0(\tau) = \frac{1}{1 + \alpha} \left(\alpha + e^{-(1+\alpha)\tau} \right).$$

However, it is clear that this solution does not at all apply for large times since a should tend to 0 when $\tau \to \infty$. It is therefore necessary to consider a different scaling, and we change the time scale from T_1 to T_2, $t^* = T_2 t$. Show that the system now becomes

$$\varepsilon \frac{dc}{dt} = -ac + \alpha(1 - c),$$
$$\frac{da}{dt} = -ca, \tag{3.38}$$

which is a *singularly* perturbed system with a small parameter in front of (one of) the highest derivatives. We also observe that $\tau = t/\varepsilon$, in accordance with system (3.36) being the "inner" system of (3.38). Inserting

$$c(t) = c_0(t) + \varepsilon c_1(t) + \cdots$$
$$a(t) = a_0(t) + \varepsilon a_1(t) + \cdots$$

gives to the leading order

$$0 = -a_0 c_0 + \alpha \left(1 - c_0\right),$$

$$\frac{da_0}{dt} = -c_0 a_0.$$

From the first equation, we obtain

$$c_0 = \frac{\alpha}{\alpha + a_0},$$

and hence,

$$\frac{da_0}{dt} = -\frac{\alpha a_0}{\alpha + a_0}.$$

This is a separable equation which may be written as

$$\left(\frac{1}{a_0} + \frac{1}{\alpha}\right) da_0 = -dt,$$

and integrated to

$$\ln a_0 + \frac{a_0}{\alpha} = -t + t_0,$$

where t_0 is an unknown constant. It is not possible to express a_0 as an elementary function of t.

We have now obtained the inner solution

$$A_0 \left(\tau\right) = 1,$$

$$C_0 \left(\tau\right) = \frac{1}{1 + \alpha} \left(1 + \alpha e^{-(1+\alpha)\tau}\right),$$

which also fulfils the initial conditions, and the outer solution,

$$c_0 \left(t\right) = \frac{\alpha}{\alpha + a_0 \left(t\right)},$$

$$\ln a_0 \left(t\right) + \frac{a_0 \left(t\right)}{\alpha} = -t + t_0,$$

with one unknown constant t_0.

The matching principle requires that

$$\lim_{\tau \to \infty} C_0 \left(\tau\right) = \lim_{t \to 0} c_0 \left(t\right),$$

$$\lim_{\tau \to \infty} A_0 \left(\tau\right) = \lim_{t \to 0} a_0 \left(t\right).$$

Since $A_0 = 1$, we need that $\lim_{t \to 0} a_0 \left(t\right) = 1$. This is fulfilled only if $t_0 = 1/\alpha$. In addition, we also have to check that

$$\frac{\alpha}{1 + \alpha} = \lim_{\tau \to \infty} C_0 \left(\tau\right) = \lim_{t \to 0} c_0 \left(t\right) = \lim_{t \to 0} \frac{\alpha}{\alpha + a_0 \left(t\right)},$$

which is true since $\lim_{t \to 0} a_0 \left(t\right) = 1$.

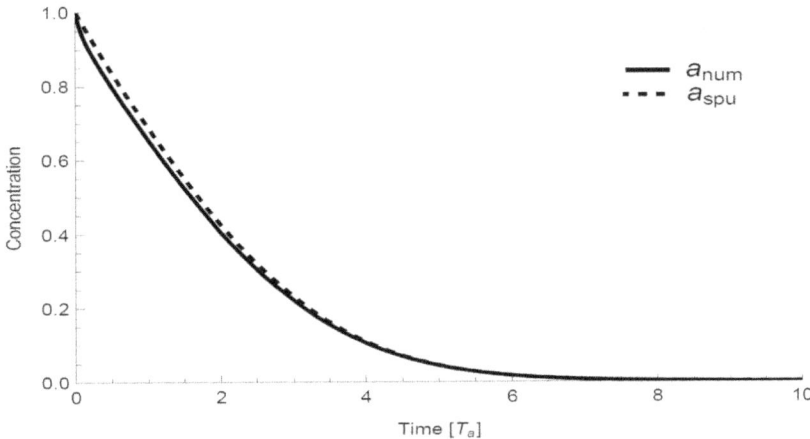

FIGURE 3.8
Numerical and uniform singular perturbation solutions for $\varepsilon = 0.1$ and $\alpha = 0.5$.

This finally leads to the uniform solution

$$a_0^u(t) = A_0\left(\frac{t}{\varepsilon}\right) + a_0(t) - \lim_{\tau \to \infty} A_0(\tau) = a_0(t),$$

$$c_0^u(t) = C_0\left(\frac{t}{\varepsilon}\right) + c_0(t) - \lim_{\tau \to \infty} c_0(t) = \frac{1}{1+\alpha}$$

$$\left(1 + \alpha e^{-(1+\alpha)t/\varepsilon}\right) + \frac{\alpha}{\alpha + a_0(t)} - \frac{\alpha}{\alpha + 1}.$$

This is not a fully explicit solution since we need to solve the transcendental equation

$$\ln a_0(t) + \frac{a_0(t) - 1}{\alpha} = -t. \tag{3.39}$$

We know that $a(0) = 1$ solves the equation, and for very small values of $a(t)$, the term $\frac{a_0(t)}{\alpha}$ is negligible compared to the logarithm. If we consider $\alpha = 1$ and require $a_0(t) = 10^{-6}$, then $t = 1 - 10^{-6} - \ln\left(10^{-6}\right) \approx 14.8$. The simple idea is now to compute a table of t-s corresponding to a_0-s in the range $\left[10^{-6}, 1\right]$, and then turn the table around and plot a_0 vs. t. Thus, we may avoid using iteration like Newton's method for solving (3.39).

Observe that the full system is nonlinear, and it is probably not possible to solve it analytically. However, the singular perturbation expansion, compared to an accurate numerical solution, shows that the uniform perturbation solution is quite accurate, see Figure 3.8. The singular perturbation solution becomes even more accurate when α increases. Further investigations are left to the reader.

3.2.5 Application II: Michaelis-Menten enzyme reaction

The differential equations of the Michaelis-Menten enzyme reaction is a famous mathematical model which admits both regular and singular perturbation analysis. It has turned out that the mathematical model also applies to several other situations quite different from the enzyme reaction setting. For the full derivation of the model, we refer to [30] or [31]. However, the analysis below deviates from these references and is based on the alternative treatment in [27].

3.2.5.1 Chemical reaction kinetics

Chemical reaction kinetics is the theory about how chemical reactions develop in time. In school we may have learned about some simple chemical reactions. For example, a fuel cell combining hydrogen and oxygen into water:

$$2H_2 + O_2 \rightarrow 2H_2O.$$

It turns out that this reaction is far too simple to explain the fuel cell operation in full, and a detailed analysis requires more than 30 interacting reactions. The core principle defining the kinetics (dynamics) of a chemical reaction is the *Law of Mass Action*, first formulated by the Norwegian scientists Cato Maximilian Guldberg and Peter Waage about 150 years ago. Somewhat simplistic, the law states that chemical reactions occur with rates that are proportional to the concentrations of the involved ingredients. The constants of proportionality are called *rate constants*. Consider the following reversible reaction where the substances A and B combine into C, which, in turn, splits into A and B:

$$A + B \underset{k_{-1}}{\overset{k_1}{\rightleftharpoons}} C.$$

Denoting the corresponding concentrations by a, b, and c, the equations based on the Law of Mass Action are

$$\frac{dc}{dt} = k_1 ab - k_{-1} c,$$

$$\frac{da}{dt} = -k_1 ab + k_{-1} c,$$

$$\frac{db}{dt} = -k_1 ab + k_{-1} c,$$

where k_1 and k_{-1} are *rate constants* (with different units – why?). The system is a nonlinear dynamic system which is simplified by observing that $d(c+a)/dt = 0$, and hence $c + a = $ constant, and similarly for C and B.

3.2.5.2 Equations and qualitative behaviour

An *enzyme* is a biological substance that enables a biochemical reaction. In real life, chemical reactions may be very complex, and the Michaelis-Menten model is often a gross simplification. The reaction is described by the following diagram

$$A + E \overset{k_1}{\underset{k_{-1}}{\rightleftharpoons}} B \overset{k_2}{\rightarrow} P + E. \tag{3.40}$$

Here A is the *substrate* which is going to be converted, E is the helping enzyme, B is an intermediate substrate/enzyme *complex*, and P is the *product* formed when the complex splits up into a new form of A and the unharmed enzyme. There is also a possibility that B turns back into A and E. Since we are later going to scale the equations, the concentrations of the various ingredients are denoted by a^*, e^*, b^*, and p^*. The model describing the above reaction is given by

$$\frac{da^*}{dt^*} = -k_1 a^* e^* + k_{-1} b^*, \tag{3.41}$$

$$\frac{de^*}{dt^*} = -k_1 a^* e^* + k_{-1} b^* + k_2 b^* \tag{3.42}$$

$$\frac{db^*}{dt^*} = k_1 a^* e^* - k_{-1} b^* - k_2 b^*, \tag{3.43}$$

$$\frac{dp^*}{dt^*} = k_2 b^*. \tag{3.44}$$

We observe by combining the equations that

$$b^* (t^*) + e^* (t^*) = \text{constant},$$
$$p^* (t^*) + a^* (t^*) + b^* (t^*) = \text{constant}.$$

Consider the following initial conditions at

$$a^* (0) = a_I, \ e^* (0) = B_M, \ b^* (0) = 0, \ p^* (0) = 0. \tag{3.45}$$

Thus,

$$e^* (t^*) = B_M - b^* (t^*), \tag{3.46}$$
$$p^* (t^*) = a_I - a^* (t^*) - b^* (t^*), \tag{3.47}$$

and we only need to consider the remaining equations for A and B. With redefined rate constants

$$k_a = k_1 B_M, \ k_r = k_2, \ k_d = k_{-1},$$

we obtain

$$\frac{da^*}{dt^*} = -k_a a^* \left(1 - \frac{b^*}{B_M} \right) + k_d b^*, \tag{3.48}$$

$$\frac{db^*}{dt^*} = k_a a^* \left(1 - \frac{b^*}{B_M} \right) - k_d b^* - k_r b^*. \tag{3.49}$$

All rate constants have now dimension $Time^{-1}$, and define three different time scales, $T_a = 1/k_a$, $T_r = 1/k_r$, and $T_d = 1/k_d$, denoted the *adsorption, reaction,* and *desorption* (opposite of adsorption) time scales, respectively. For an efficient process, it is reasonable to assume that $T_a \ll T_r \le T_d$. Thus, the formation of the complex is fast, whereas the disintegration into product and enzyme, or substrate and enzyme, is considerably slower.

Clearly,

$$0 \le a^*, \ p^* \le a_I,$$

and since adding the remaining equations leads to

$$\frac{\mathrm{d}\,(a^* + b^*)}{\mathrm{d}t^*} = -k_r b^* \le 0, \qquad\qquad (3.50)$$

$a^* + b^*$ never exceeds its initial value, a_I. Similarly, there is no way $\mathrm{d}b^*/\mathrm{d}t^*$ could remain positive when b^* reaches B_M. Thus,

$$0 \le b^*\,(t) \le \min\,(a_I, B_M)\,. \qquad\qquad (3.51)$$

The qualitative behaviour of the solutions of the system (3.48)–(3.49), with the initial conditions in (3.45), is easy to envisage. The amount a^* decreases from a_I to 0 as the time increases, whereas b^* at first increases and then decays towards 0 as the substrate becomes depleted.

Time scales are defined by the reaction rates, and the obvious scale for a^* is a_I. As long as $B_M \le \mathcal{O}\,(a_I)$, it is also reasonable to use B_M as a scale for b^*. However, should a_I be much less than B_M, the appropriate scale for b^*, in accordance with (3.51), is a_I.

The reaction starts with an initial loading of the enzyme forming the complex and lasting $\mathcal{O}\,(T_a)$, during which b^* reaches its maximum value. If $a_I \ll B_M$, the term $(1 - b^*/B_M)$ will always stay close to 1 and the solution is almost independent of B_M. On the other hand, if $B_M \ll a_I$, the kinematics is effectively slowed down by the converting capacity of the enzyme. After the initial phase, where the enzyme is charged up to the limit B_M, there is now a *quasi stationary* (QS) stage where $b^* \approx B_M$ and the reaction is governed by Eq. 3.50, implying an approximate linear decay in a^*. In the final stages of the reaction, all substrate has been consumed, and both complex and substrate decay to 0 and p^* approaches its asymptotic value.

3.2.5.3   Scaling and perturbation analysis

The problem has three dimensionless parameters, which may be chosen as the ratio B_M/a_I, also used in [30], and two time scale ratios, e.g., T_a/T_r and T_r/T_d. The ratio B_M/a_I could in principle be both large and small, whereas T_a/T_r is assumed to be small, and T_r/T_d is $\mathcal{O}\,(1)$, or somewhat smaller. The following notation will be used below:

$$\kappa = \frac{B_M}{a_I}, \quad \varepsilon = \frac{T_a}{T_r}, \quad \gamma = \frac{T_r}{T_d}.$$

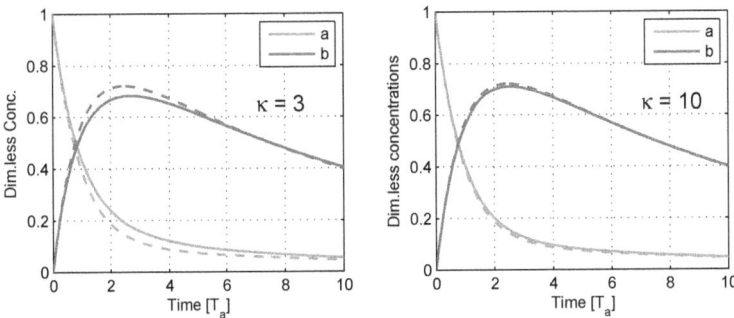

FIGURE 3.9
Numerical (dashed and solid) and leading order regular perturbation (dotted and dotdashed) expansion in $1/\kappa$ for $\kappa > 1$. Other parameters: $\gamma = 1$, $\varepsilon = 0.1$. Note that the expansion is in terms of $1/\kappa$ and not ε.

It is convenient to analyse different cases according to the size of κ.

We first consider the case where $1 \ll \kappa$ ($a_I \ll B_M$). Since $b^* \leq a_I$, it is now reasonable to use a_I as a scale both for a^* and b^*. Along with $\tau = t^*/T_a$, the dimensionless equations become

$$\frac{da}{d\tau} = -a\left(1 - \frac{1}{\kappa}b\right) + \varepsilon\gamma b, \qquad (3.52)$$

$$\frac{db}{d\tau} = a\left(1 - \frac{1}{\kappa}b\right) - \varepsilon(1+\gamma)b. \qquad (3.53)$$

The small parameter $1/\kappa$ turns this into a regular perturbation problem with linear leading order equations. The corresponding system matrix has, if $\gamma > 0$, two different negative eigenvalues, in accordance with a solution dying out as $\tau \to \infty$. The full solutions of (3.52)–(3.53) converge nicely on $[0, \infty)$ to the leading order solution when $1/\kappa \to 0$. This is illustrated in Figure 3.9. When κ is smaller than 1, the scaling is changed to $a^* = a_I a$, $b^* = B_M b$, whereas $\tau = t^*/T_a$ is now the fast dimensionless time for the initial development. Inserted into (3.48)–(3.49), we obtain

$$\frac{da}{d\tau} = -a(1-b) + \kappa\varepsilon\gamma b, \qquad (3.54)$$

$$\kappa\frac{db}{d\tau} = a(1-b) - \kappa\varepsilon b(\gamma+1). \qquad (3.55)$$

Since ε is assumed to be small, this becomes a regular perturbation system with respect to ε. Writing $A_0(\tau) + \varepsilon A_1(\tau) + \cdots$ and $B_0(\tau) + \varepsilon B_1(\tau) + \cdots$, for the inner perturbation expansions, we obtain to leading order,

$$\frac{dA_0}{d\tau} = -A_0 \left(1 - B_0\right),$$ (3.56)

$$\kappa \frac{dB_0}{d\tau} = A_0 \left(1 - B_0\right),$$ (3.57)

$$A_0 \left(0\right) = 1, \ B_0 \left(0\right) = 0.$$ (3.58)

The system is easily solved by first adding the equations to obtain

$$B_0 \left(\tau\right) = \frac{1 - A_0 \left(\tau\right)}{\kappa},$$

and then inserting B_0 into (3.56), leading to the logistic equation

$$\frac{dA_0}{d\tau} = \frac{A_0}{\kappa} \left[1 - \kappa - A_0\right],$$

with solution

$$A_0 \left(\tau\right) = \begin{cases} \frac{1-\kappa}{1-\kappa \exp\left[\frac{\tau}{\kappa}(\kappa-1)\right]}, & \kappa \neq 1, \\ \frac{1}{\tau+1}, & \kappa = 1. \end{cases}$$

For the later matching to the outer solution, we observe that

$$\lim_{\tau \to \infty} A_0 \left(\tau\right) = \max \left(0, 1 - \kappa\right),$$

$$\lim_{\tau \to \infty} B_0 \left(\tau\right) = \min \left(1, 1/\kappa\right).$$

Switching to the *slow*, outer time scale, $t^* = T_r t$, $t = \varepsilon \tau$, the system turns into

$$\varepsilon \frac{da}{dt} = -a \left(1 - b\right) + \kappa \varepsilon \gamma b,$$ (3.59)

$$\varepsilon \kappa \frac{db}{dt} = a \left(1 - b\right) - \kappa \varepsilon b \left(\gamma + 1\right).$$ (3.60)

The leading order outer equation now becomes

$$a_0 \left(1 - b_0\right) = 0,$$

which requires $a_0 \left(t\right) = 0$ or $b_0 \left(t\right) = 1$. Since κ is less than 1, $B_0 \left(\infty\right) = 1$ matches $b_0 \left(t\right) \equiv 1$. Then a_0 is free to vary and by adding (3.59)–(3.60), we obtain the (exact) dimensionless equation

$$\frac{d}{dt} \left(a + \kappa b\right) = -\kappa b.$$ (3.61)

By applying (3.61) with $b_0 = 1$ and $A_0 \left(\infty\right) = 1 - \kappa$, the matched outer

solution becomes

$$a_0 \left(t\right) = \left(1 - \kappa\right) - \kappa t.$$

The outer solution $b_0(t) = 1$ shows a nearly constant plateau for the time interval where $a_0(t)$ decreases from $1 - \kappa$ to $O(\varepsilon)$. The length, t_P, of the plateau is estimated from

$$a_0(t_P) = (1 - \kappa) - \kappa t_P = 0,$$

that is,

$$t_P = \frac{1 - \kappa}{\kappa}. \tag{3.62}$$

However, the outer solution does not have the correct behaviour for large times since it does not converge to 0. Since no further matching can be carried out, the singular perturbation analysis is not able to provide a uniform approximation on the full interval from 0 to ∞.

A simple fix, in accordance with (3.61), would be to define

$$a_0(t) = \begin{cases} (1 - \kappa) - \kappa t, & t \leq t_p, \\ 0, & t_p < t, \end{cases}$$

and

$$b_0(t) = \begin{cases} 1, & t \leq t_p, \\ e^{-(t-t_p)}, & t_p < t. \end{cases}$$

By constructing the singular perturbation matched approximations and incorporating the correction for the asymptotic behaviour in the terminal stage of the reaction, we obtain

$$a_0^{(u)}(t) = \left[A_0\left(\frac{t}{\varepsilon}\right) - \kappa t \right]^+, \tag{3.63}$$

$$b_0^{(u)}(t) = \begin{cases} \frac{1 - A_0\left(\frac{t}{\varepsilon}\right)}{\kappa}, & t \leq t_p, \\ e^{-(\frac{t}{\varepsilon}-t_p)}, & t_p < t, \ t_P = \frac{1-\kappa}{\kappa}. \end{cases} \tag{3.64}$$

The graphs below show $a(t)$ and $b(t)$ from numerical simulations (solid lines), the leading order uniform solutions, stated in (3.63)–(3.64) (dashed lines), and the product $p(t)$ computed from

$$p(t) = 1 - a(t) - b(t).$$

Dimensionless variables are used throughout; b is re-scaled in the plot by $1/\kappa$, so that both a, b, and p range between 0 and 1. Figure 3.10 shows what happens for $\kappa = 0.1$ and a moderately small value 0.1 for $\varepsilon = 0.1$. The initial transient dies fast, and the quasi-stationary plateau forms. The estimate of its length in Eq. 3.62, $t_p = (1 - \kappa)/\varepsilon \kappa = 90$, is reasonable, although the uniform solutions are not very accurate near the end of the plateau. By further decreasing ε to 0.01, the plateau is extended by a factor of 10 compared to the solution shown in Figure 3.10. The solutions are shown in Figure 3.11, where the smaller ε, as expected, gives a much better agreement between the numerical and approximate solutions. Even if the approximate solutions

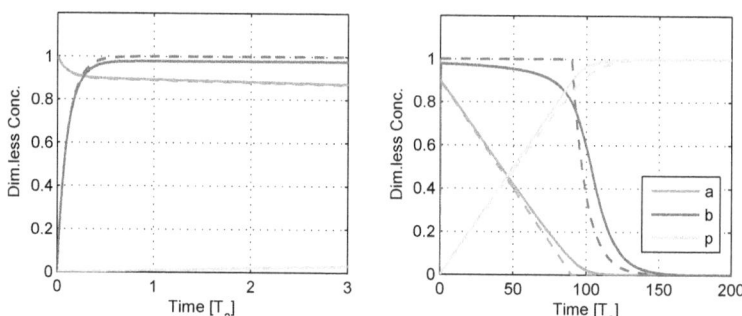

FIGURE 3.10
Uniform approximation (dashed) and exact numerical solution (solid) with $\kappa = 0.1$ and $\varepsilon = 0.1$, $\gamma = 1$. The left plot shows an expanded view of the initial behaviour.

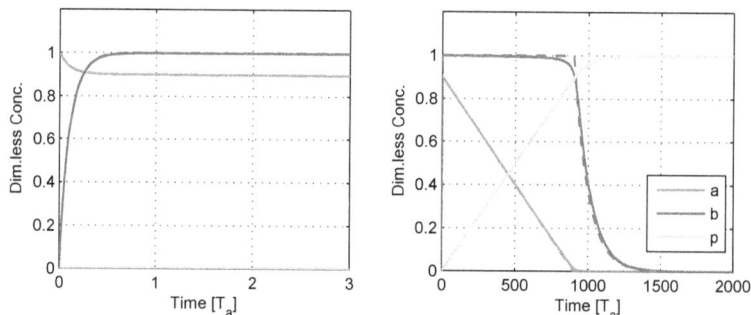

FIGURE 3.11
Similar to Figure 3.10 with ε reduced by a factor of 10, $\varepsilon = 0.01$.

look reasonable, the simple remedy above for extending the solution past the plateau does not work well for κ slightly less than 1.

An expanded discussion of the Michaelis-Menten equations is found in [27]. It should be observed that the singular perturbation analysis presented in [30] and [31] is not consistent with the assumptions above and, contrary to what is claimed in [30], does not easily show the quasi-stationary plateau observed here.

3.2.6   Exercises

1. Apply the dominant balancing method to determine the leading order behaviour of the roots of the following algebraic equations $(0 < \varepsilon \ll 1)$:

(a) $\varepsilon x^4 + \varepsilon x^3 - x^2 + 2x - 1 = 0$.

(b) $\varepsilon x^3 + x - 2 = 0$.

(c) Solve the equation exactly by applying some symbolic manipulation software, *e.g.*, Maple^{TM}. Compare the answers.

2. Use singular perturbation methods to obtain a uniform approximate solution to the following problems. In each case assume $0 < \varepsilon \ll 1$ and $0 < x < 1$.

(a) $\varepsilon y'' + 2y' + y = 0$, $y(0) = 0$, $y(1) = 1$.

(b) $\varepsilon y'' + y'^2 = 0$, $y(0) = \frac{1}{4}$, $y(1) = \frac{1}{2}$.

(c) $\varepsilon y'' + (1 + x)y' + y = 1$, $y(0) = 0$, $y(1) = 1$.

(d) $\varepsilon y'' + 2y'^y = 1$, $y(0) = y(1)$.

3. Examine the exact solution to show why the singular perturbation method fails on the boundary value problem

$$\varepsilon u'' + u = 0, \; 1 \le x \le 2,$$
$$u(1) = 1, \; u(2) = 2.$$

4. Obtain a uniform approximation solution to the problem

$$\varepsilon u'' - (2x + 1)u' + 2u = 0, \; 0 < x < 1,$$
$$u(0) = 1, \; u(1) = 0.$$

5. Is the problem

$$\varepsilon y'' + \frac{1}{x}y' + y = 0, \; x > 0,$$
$$y(0) = 1, \; y'(0) = 0,$$

a singular perturbation problem? Discuss.

6. Show that the regular perturbation fails on the boundary value problem

$$\varepsilon y'' + y' + y = 0, \; 0 < x < 1, \; 0 < \varepsilon \ll 1,$$
$$y(0) = 0, \; y(1) = 1.$$

Find the exact solution and sketch it for $\varepsilon = 0.05$ and $\varepsilon = 0.005$. If $x = \mathcal{O}(\varepsilon)$, show that $\varepsilon y''(x)$ is large; if $x = \mathcal{O}(1)$, show that $\varepsilon y''(x) = \mathcal{O}(1)$. Find the inner and outer approximations from the exact solution.

7. The density of cells, n^*, in a part of the body may be modelled as

$$\frac{dn^*}{dt^*} = \alpha n^* - \omega n^*, \tag{3.65}$$

where α is the birth rate and ω the death rate of the cells. In order to prevent that production of cells runs astray, normal cells produce a chemical called an *inhibitor*, which dampens uncontrolled growth. The inhibitor has density c^* and works by changing the birth rate to

$$\alpha = \frac{\alpha_0}{1 + c^*/A}.$$

The production of the inhibitor is proportional to n^*, whereas the destruction of the inhibitor is proportional to its density,

$$\frac{dc^*}{dt^*} = \beta n^* - \delta c^*. \tag{3.66}$$

The system consisting of the equations 3.65 and 3.66 has a time scale ω^{-1} connected to the breakdown of the cells, and a time scale δ^{-1} connected to the breakdown of the inhibitor. It is known that $\omega^{-1} \gg \delta^{-1}$.

(a) Scale the system by applying ω^{-1} as the time scale and A as a scale for c^*. Show that the system with a certain scale for n^* then may be written

$$\dot{n} = \left(\frac{\kappa}{1 + c} - 1 \right) n,$$

$$\varepsilon \dot{c} = n - c. \tag{3.67}$$

What is the meaning of ε and κ? What can be said about the size of ε, and what is such a system called? Determine what kind of equilibrium point the trivial equilibrium point $(0, 0)$ is.

(b) Determine the orbit and the equation for the motion of the outer solution of Eq. (3.67) to leading order $(O(1))$. Show, without necessarily solving the differential equation, that all motion on this path converges to an equilibrium point for the full system.

(c) Determine, to leading order, the inner solution of (3.67) by introducing a new time scale. Finally state a uniform, approximate solution (it is not possible to solve the equation in (b) explicitly).

Note: For (b) and (c) we assume that κ is larger than 1.

4

Stability Analysis

4.1 One-dimensional Flow

First order, one-dimensional autonomous ordinary differential equations which arise in many applications, have the form

$$\frac{\mathrm{d}u\,(t)}{\mathrm{d}t} = f\,(u\,(t))\,, \tag{4.1}$$

where $u = u(t)$ is the dependent variable of interest, t the time, and f a given function describing the time evolution of u. Since (4.1) is separable, assuming f to be continuous, the differential equation (4.1) can be reduced to the integral equation

$$\int \frac{\mathrm{d}u}{f\,(u)} = t + C.$$

However, the integral can be solved analytically only for a very few forms of f and may not result in explicit solutions. Instead of looking for or approximating the analytic solutions, this chapter focuses on the qualitative features of solutions. A key to the qualitative analysis is to first determine the *equilibrium points* (steady states corresponding to constant solutions), then the dynamics of solutions close to the equilibria.

Equilibrium points of the system (4.1) are real solutions of the algebraic equation

$$f(u) = 0.$$

We shall, for simplicity, assume that f is a *continuous* function with *isolated zeros*. This means that if u^* is an equilibrium point, $f\,(u^*) = 0$, then there is a *neighbourhood* around u^* (an open interval containing u^*) which contains no other equilibrium points.

Figure 4.1 shows an example of a function $f(u)$ having simple zeroes at $u = 1$ and 2, and a double zero at $u = 3$. Between the equilibria we find intervals for which the solutions $u\,(t)$ are *strictly increasing*, $u'\,(t) = f\,(u) > 0$, or *decreasing*, $u'\,(t) = f\,(u) < 0$. The direction of the flow of $u\,(t)$ as time increases may be indicated by arrows on the u-axis, also called the *phase line*. A strictly increasing function will either converge to a finite limit, or

DOI: 10.1201/9781003725206-4

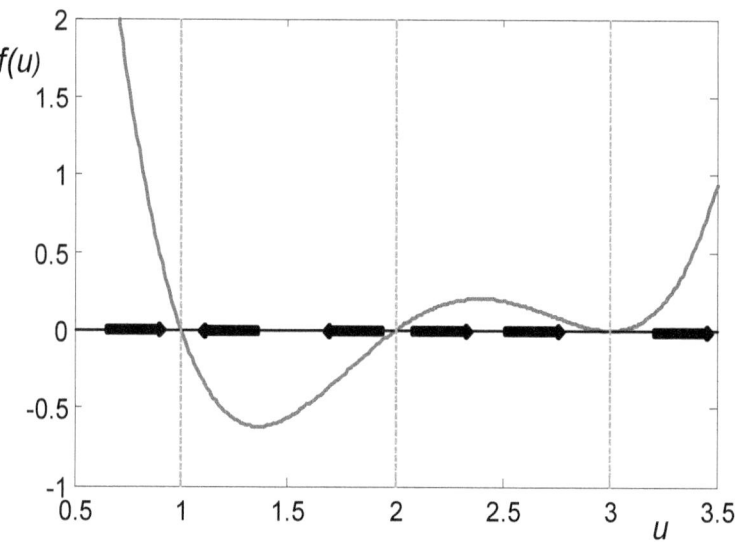

FIGURE 4.1
Example of a right hand side function showing three equilibrium points. The arrows indicate the direction of the motion of u if one starts in between the equilibrium points.

diverge to $+\infty$, similarly for strictly decreasing functions. In the present case, the convergence will thus be towards an equilibrium point or to $+\infty$ or $-\infty$. Figure 4.2 illustrates three of the cases that may occur for isolated equilibrium points (the 4th case is a mirror image of $u^* = 3$). Solutions near the equilibrium point $u^* = 1$ converge towards it, and the point $u^* = 1$ is called an *attractor*. Similarly, the equilibrium point $u^* = 2$ is called a *repeller*, whereas the point $u^* = 3$, where solutions converge from one side and move away from the other, has no common name. Repellers and the point $u^* = 3$ in Figure 4.1 are *unstable* because small perturbations from the equilibrium may cause the system to go to a different equilibrium or even diverge to infinity. When a solution returns to the steady state u^* for all reasonable perturbations, no matter how large they are, then the equilibrium point u^* is *globally asymptotically stable*.

In general, we have the following result.

Theorem 4.1 *An equilibrium point u^* of the system (4.1) is said to be stable if $f'(u^*) < 0$, and unstable if $f'(u^*) > 0$.*

The case where $f'(u^*) = 0$ requires further investigation.

Example 4.1 *The logistic growth model (see the following chapter on population models)*

$$\frac{du}{dt} = ru\left(1 - \frac{u}{k}\right), \quad \text{with } r, k > 0,$$

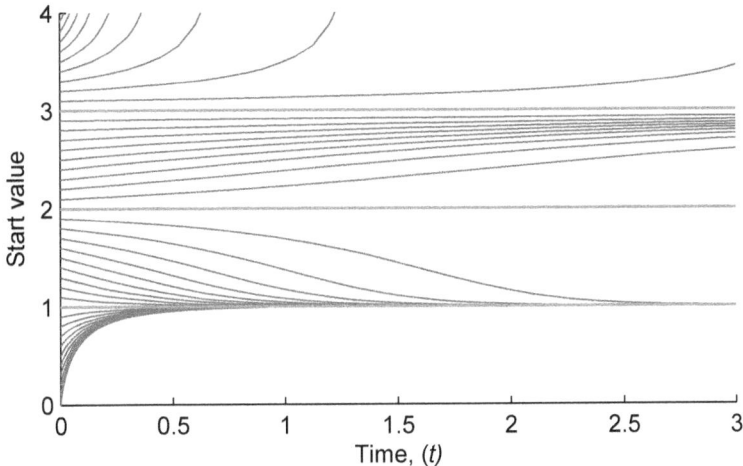

FIGURE 4.2
Numerical solutions of the equation (4.1) corresponding to $f(u)$ in Figure 4.1, for different initial conditions. All solutions for which $u(0) < 2$ converge to 1. When $2 < u(0) < 3$, the solutions converge to 3. Finally, for $3 < u(0)$, the solutions diverge to ∞.

has two equilibrium points $u^* = 0$ *which is unstable* $(f'(0) = r > 0)$ *and* $u^* = k$ *which is stable* $(f'(k) = -r < 0)$.

The concept of stability is very important in modelling. Consider a population of fish in a lake being in a locally asymptotically stable state u^*. Toxic chemicals dumped into the lake may cause the population of fish to drop. Local asymptotic stability means that for a small drop, the system will return to the original state u^* over time. If many fish are killed in the event, then the perturbation is not small and there is no guarantee that the fish population will return to its original state. On the contrary, it may approach some other equilibrium state or even become extinct. We will present a general framework to investigate stability.

4.2 Higher-dimensional Flow

We now consider an initial-value problem (IVP) of the form

$$\dot{x}(t) = f(x(t)),$$
$$x(t_0) = x_0,$$

(4.2)

where $x = (x_1, \ldots, x_n) \in \Omega \subset \mathbb{R}^n$, Ω open, bounded, and connected, $f : \Omega \longrightarrow \mathbb{R}^n$ a mapping, $x_0 \in \Omega$ a constant vector, and $t \in I \subset \mathbb{R}$ (I an interval containing t_0). Before studying the solutions of the IVP, it is important to ensure that the problem is well-posed.

An initial value problem is said to be mathematically well-posed (in the sense of Hadamard) if the following conditions hold:

- Its solution exists,

- its solution is unique,

- its solution continuously depends on the initial conditions.

The following theorem is a well-known result on the well-posedness of the system (4.2).

Theorem 4.2 *(Picard–Lindelöf theorem) [11]*
Assume that for every $x \in \Omega$, there exist $\delta > 0$ and $c > 0$ such that

$$\|f(y) - f(z)\| \leq c\left(\|y - z\|\right), \quad \|f(y)\| \leq c,$$

for every $y, z \in B(x, \delta)$. Then, for any $x_0 \in \Omega$, the solution of the Cauchy problem (4.2) exists locally in the interval $(t_0 - \epsilon, t_0 + \epsilon)$, with $\epsilon > 0$. Moreover, the solution is unique.

The proof of the above theorem can be found in [37]. The existence is done by applying the method of successive approximation to construct a sequence of uniformly bounded equicontinuous functions, which has a subsequence that converges to the solution of the IVP. The uniqueness is done by applying Gronwall's lemma. A corollary to the above theorem is the Cauchy-Lipschitz theorem, which requires the function f to be of class C^1.

4.2.1   Definitions

Definition 4.1 *An equilibrium point (or a steady state) x^* of the system (4.2) is a real solution of the algebraic system of equations*

$$f(x) = 0.$$

Definition 4.2

(i.) An equilibrium point x^* is said to be Lyapunov stable if for all $\epsilon > 0$, there exists $\delta > 0$ such that if

$$\|x(0) - x^*\| < \delta, \quad then \quad \|x(t) - x\| < \epsilon.$$

(ii.) An equilibrium point x^* is said to be asymptotically stable if it is Lyapunov stable and there exists $\delta > 0$ such that if

$$\|x(0) - x^*\| < \delta, \quad then \quad \lim_{t \to \infty} \|x(t) - x^*\| = 0.$$

(iii.) An equilibrium point x^* is said to be exponentially stable if there exists $\alpha > 0, \beta > 0, \delta > 0$ such that if

$$\|x(0) - x\| < \delta, \quad then \quad \|x(t) - x^*\| < \alpha\|x(0) - x^*\|e^{-\beta t} \quad for \ all \quad t \geq 0.$$

(iv.) An equilibrium point x^* is said to be globally stable if $\lim\limits_{t \to \infty} x(t) = x^*$ for all $t \geq 0$.

(v.) An equilibrium point that is not stable is called unstable.

4.2.2 Linear systems

In this section, we assume that the system (4.2) is in the form

$$\frac{dx}{dt} = \mathcal{A}x, \quad x(0) = x_0, \tag{4.3}$$

where $\mathcal{A} = (a_{i,j})_{i,j=1,\ldots,n} \in \mathcal{M}(n \times n)$ is a matrix with constant coefficients.

Proposition 4.1

i. If all the eigenvalues of $\mathcal{A}$ have negative real part, then given any solution $x(t)$ of (4.3), there exist positive constants M and b such that

$$\|x(t)\| \leq Me^{-bt}, \quad \forall t > 0$$

and hence

$$\lim\limits_{t \to \infty} \|x(t)\| = 0.$$

In this case, the equilibrium point $x^* = \mathbf{0}$ is asymptotically stable.

ii. If there exists an eigenvalue of $\mathcal{A}$ with a positive real part, the equilibrium point $x^* = \mathbf{0}$ is unstable.

We now consider the special case where $n = 2$, the 2-dimensional autonomous linear system [14]

$$\begin{pmatrix} x'(t) \\ y'(t) \end{pmatrix} = \mathcal{A}\begin{pmatrix} x(t) \\ y(t) \end{pmatrix}, \quad \mathcal{A} = \begin{pmatrix} a & b \\ c & d \end{pmatrix}. \tag{4.4}$$

Let λ_1 and λ_2 be the eigenvalues of the matrix $\mathcal{A}$.

Case 1: Assume $\lambda_1 \neq \lambda_2$. Then, $\mathcal{A}$ is diagonalizable and can be written in the form $\mathcal{A} = PDP^{-1}$, with P an invertible matrix formed by the eigenvectors associated with the eigenvalues, and D the diagonal matrix formed by the eigenvalues λ_1 and λ_2. Using the transformation

$$\begin{pmatrix} u \\ v \end{pmatrix} = P\begin{pmatrix} x \\ y \end{pmatrix},$$

the system (4.4) is reduced to the decoupled system

$$\begin{pmatrix} u' \\ v' \end{pmatrix} = D \begin{pmatrix} u \\ v \end{pmatrix} = \begin{pmatrix} \lambda_1 & 0 \\ 0 & \lambda_2 \end{pmatrix} \begin{pmatrix} u \\ v \end{pmatrix}. \tag{4.5}$$

This results in the solution

$$u(t) = u(0)e^{\lambda_1 t}, v(t) = v(0)e^{\lambda_2 t},$$

and the relation

$$v = k u^\lambda,$$

where k is a constant, and $\gamma = \dfrac{\lambda_2}{\lambda_1}$. We have the following results:

a. If $\lambda_1, \lambda_2 \in \mathbb{R} \setminus \{0\}$ and have same signs, then the origin $(0,0)$ is a node. It is stable when $\lambda_1 < 0$ and unstable when $\lambda_1 > 0$. The phase portrait of solution is given in Figure 4.3(a).

b. If $\lambda_1, \lambda_2 \in \mathbb{R}$ and have opposite signs, the origin $(0,0)$ is unstable. It is a saddle point. See the phase portrait in Figure 4.3(b).

c. If $\lambda_1 = 0$ or $\lambda_2 = 0$, then the origin is a non-isolated equilibrium point; we have a whole line of equilibrium points.

d. If $\lambda_1, \lambda_2 \in \mathbb{C}$, with $\lambda_1 = \alpha + i\beta$. Assume u of the form $u = re^{i\theta}$ and substitute into (4.5), then we obtain

$$r'(t) = \alpha r, \quad \theta'(t) = \beta. \tag{4.6}$$

This leads to the solution

$$r(t) = r_0 e^{\alpha t}, \theta(t) = \beta t + \theta_0,$$

and the relation

$$r(\theta) = r_0 e^{\alpha \left(\frac{\theta - \theta_0}{\beta} \right)} = ae^{b\theta}. \tag{4.7}$$

    i. If $\alpha \neq 0$, (4.7) is the equation of a logarithmic spiral. Therefore, the origin $(0,0)$ is a spiral. It is a stable when $\alpha < 0$, and an unstable when $\alpha > 0$.

    ii. If $\alpha = 0$, the radius in (4.6) is constant and the origin is a centre.

The direction of flow is counter-clockwise when $\beta > 0$ and clockwise when $\beta < 0$. The phase portrait is illustrated in Figure 4.3(c).

Case 2: Assume $\lambda_1 = \lambda_2 = \lambda$.

a. If there are two independent eigenvectors associated with λ, then $(0,0)$ is a star. It is stable when $\lambda < 0$ and unstable when $\lambda > 0$ (see Figure 4.4(a)).

b. If there is only one eigenvector associated with λ, $(0,0)$ is a degenerate node. It is stable when $\lambda < 0$ and unstable when $\lambda > 0$ (see Figure 4.4(b)).

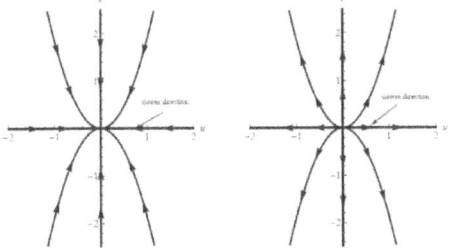

(a) Stable and unstable nodes

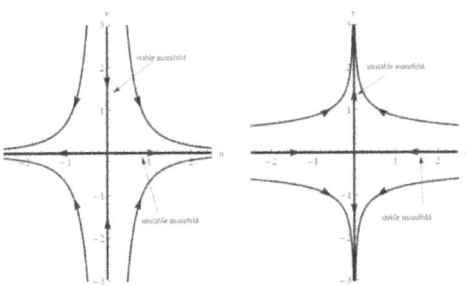

(b) Saddle points

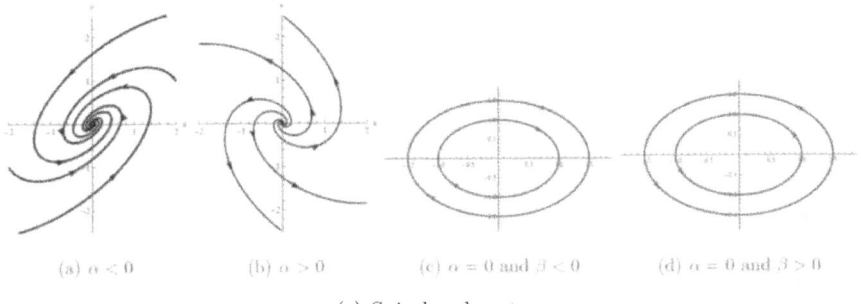

(a) $\alpha < 0$ (b) $\alpha > 0$ (c) $\alpha = 0$ and $\beta < 0$ (d) $\alpha = 0$ and $\beta > 0$

(c) Spiral and centre

FIGURE 4.3
Phase portrait of solutions when the eigenvalues are distinct.

4.2.3   Nonlinear systems

We now assume that the function $f = (f_1, \ldots, f_n)$ in (4.2) is nonlinear and of class C^1.

Let $x^* = (x_1^*, \ldots, x_n^*)$ be an equilibrium point of the system (4.2). Consider a small perturbation $u = x - x^*$ near x^*. By the Taylor's expansion, we have

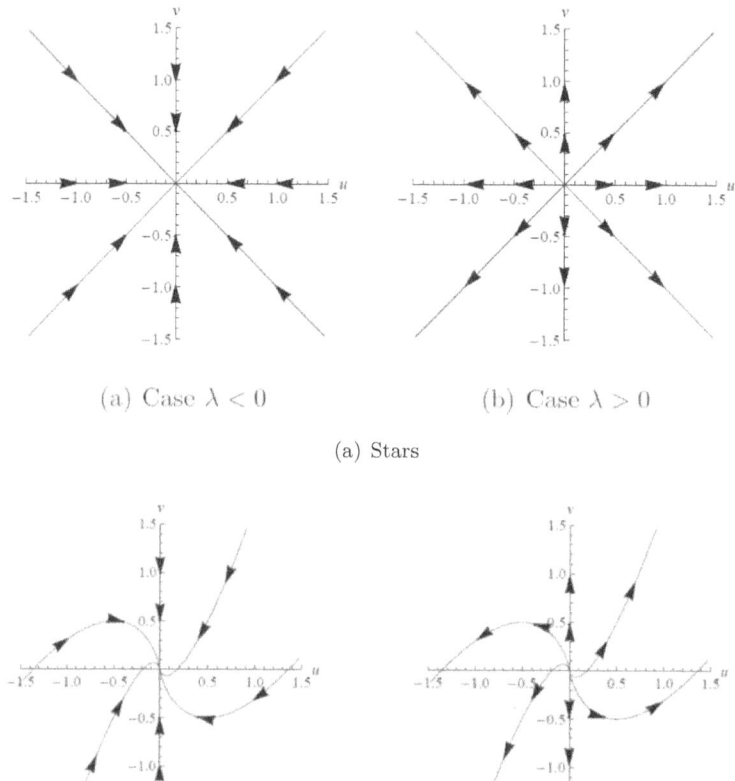

(a) Case $\lambda < 0$ (b) Case $\lambda > 0$

(a) Stars

(b) Degenerate nodes

FIGURE 4.4
Phase portrait of solutions when there are repeated eigenvalues.

for $i \in \{1, \ldots, n\}$:

$$
\begin{aligned}
u_i'(t) &= x_i'(t) \\
&= f_i(u + x^*) = f_i(u_1 + x_1^*, \ldots, u_n + x_n^*) \\
&= u_1 \frac{\partial f_i}{\partial x_1}(x^*) + \ldots u_n \frac{\partial f_i}{\partial x_n}(x^*) + \mathcal{O}(|u|^2).
\end{aligned}
$$

In the matrix form, the above perturbed system can be written as

$$
u'(t) = \mathcal{J}u + \mathcal{O}(|u|^2),
$$

where $\mathcal{J}$ is the Jacobian matrix of the system (4.2) evaluated at x^*, given by

$$\mathcal{J} = \begin{pmatrix} \frac{\partial f_1}{\partial x_1}(x^*) & \cdots & \frac{\partial f_1}{\partial x_2}(x^*) \\ \vdots & & \vdots \\ \frac{\partial f_n}{\partial x_1}(x^*) & \cdots & \frac{\partial f_n}{\partial x_1}(x^*) \end{pmatrix}.$$

The system

$$u'(t) = \mathcal{J}u$$

is called the *linearized system* associated with (4.2) near x^*. Moreover, if the real part of all the eigenvalues of the matrix $\mathcal{J}$ is nonzero, then x^* is called a *hyperbolic* equilibrium point.

Theorem 4.3 (Hartman-Grobman theorem [21, 23])
*If the system (4.2) has a hyperbolic equilibrium point x^*, then in a neighborhood of x^*, the system (4.2) and its associated linearized system $u' = Au$ are **topological conjugate** (there exists a homeomorphism that will conjugate one into the other).*

4.2.4   Routh-Hurwitz criteria

We have seen that the stability depends on the signs of the eigenvalues. It is difficult to explicitly find the eigenvalues (roots of a polynomial of degree n), especially when the dimension of the system is higher ($n > 2$). In such a situation, the following Routh-Hurwitz criteria for stability can be applied [4].

Consider the (characteristic) equation of degree n

$$p(\lambda) = a_0\lambda^n + a_1\lambda^{n-1} + \ldots + a_{n-1}\lambda + a_n,$$

where the $a_i's$, $i = 1, ..., n$, are real constant coefficients and $a_0 > 0$. Define the n Hurwitz matrices:

$$H_1 = (a_1), H_2 = \begin{pmatrix} a_1 & a_0 \\ a_3 & a_2 \end{pmatrix}, H_3 = \begin{pmatrix} a_1 & a_0 & 0 \\ a_3 & a_2 & a_1 \\ a_5 & a_4 & a_3 \end{pmatrix},$$

$$H_n = \begin{pmatrix} a_1 & a_0 & 0 & \vdots & 0 \\ a_3 & a_2 & a_1 & \vdots & 0 \\ a_5 & a_4 & a_3 & \vdots & 0 \\ \cdots & \cdots & \cdots & \cdots & \cdots \\ 0 & 0 & 0 & \vdots & a_n \end{pmatrix}.$$

All the roots of the polynomial $p(\lambda)$ have negative real parts if and only if $a_0 > 0$ and the determinants of the n Hurwitz matrices are positive, that is,

$$\Delta_1 = a_1 > 0, \quad \Delta_2 = \det(H_2) > 0, \cdots, \Delta_n = \det(H_n) > 0. \tag{4.8}$$

In particular,

i. When $n = 1$, conditions (4.8) reduce to $a_1 > 0$.

ii. When $n = 2$, conditions (4.8) reduce to $a_1 > 0$ and $a_2 > 0$.

iii. When $n = 3$, conditions (4.8) reduce to $a_1 > 0$, $a_2 > 0$, $a_3 > 0$ and $a_1 a_2 - a_0 a_3 > 0$.

iv. When $n = 4$, conditions (4.8) reduce to $a_1 > 0$, $a_2 > 0$, $a_3 > 0$, $a_4 > 0$, $a_1 a_2 - a_0 a_3 > 0$ and $a_1 a_2 a_3 - a_1^2 a_4 - a_0 a_3^2 > 0$.

v. When $n = 5$, conditions (4.8) reduce to $a_1 > 0$, $a_2 > 0$, $a_3 > 0$, $a_4 > 0$, $a_5 > 0$, $a_1 a_2 - a_0 a_3 > 0$, $a_1 a_2 a_3 - a_1^2 a_4 - a_0 a_3^2 > 0$ and $a_1 a_3 a_5 - a_5^2 - a_1 a_2^2 a_5 - a_2 a_3^2 a_5 + a_1 a_4 a_5 - a_1^2 a_4^2 + a_1 a_2 a_3 a_4 - a_3^2 a_4 > 0$.

4.2.5 Lyapunov theorem

Theorem 4.4 *(Lyapunov theorem for global stability) [25]*
Let x^* be a steady state for the system (4.2). Assume that we can find a Lyapunov function, i.e., a continuously differentiable, real-valued function $V(x)$ with the following properties:

i. $V(x) > 0$ *for* $x^* \neq x$,

ii. $V(x^*) = 0$,

ii. $\frac{dV}{dt} < 0$ *for* $x^* \neq x$.

Then, x^* is globally asymptotically stable.

4.3 Hopf Bifurcation and Limit Cycle

A system's qualitative behaviour may alter as its parameters change. Equilibrium points can be created or destroyed, or their stability can change. Bifurcations are these qualitative shifts in the dynamics, and the parameter values at which they occur are called bifurcation points or bifurcation parameters.

A Hopf bifurcation is a type of bifurcation that occurs when two complex conjugate eigenvalues cross the imaginary axis of the complex plane, causing an equilibrium point to lose stability. Hence, a stable spiral switches into an unstable spiral. Supercritical and subcritical Hopf bifurcations are the two different forms of Hopf bifurcations. The distinction is in the equilibrium point's stability at the bifurcation point. The Hopf bifurcation is supercritical if the equilibrium point at the bifurcation point is stable; otherwise, it is subcritical.

The occurrence of Hopf bifurcation in a system informs the presence of limit cycles. A limit cycle is an isolated closed trajectory. Isolated in the sense that neighboring trajectories are not closed. It has been demonstrated by Poincaré, Andronov, and Hopf that in the case of a supercritical Hopf bifurcation, a limit cycle always happens after the bifurcation point; in the case of a subcritical Hopf bifurcation, it occurs before the bifurcation point. The following theorem summarizes their findings.

Theorem 4.5 *(Poincaré, Andronov and Hopf theorem [see Moiola and Chen (1996)])*
Consider the two-dimensional system with parameter μ

$$\frac{dx}{dt} = f(x, y, \mu),$$
$$\frac{dy}{dt} = g(x, y, \mu).$$

Suppose that (x^*, y^*) is an equilibrium point and exists for all values of μ. Suppose also that the eigenvalues of the Jacobian matrix evaluated at (x^*, y^*) is of the form $\lambda_{1,2} = \alpha(\mu) \pm i\beta(\mu)$ and cross the imaginary axis (in the complex plane) at $\mu = \mu_0$ (i.e., $\alpha(\mu_0) = 0$).

i. *If at $\mu = \mu_0$ the nonlinear system is stable around (x^*, y^*), then the system has a limit cycle for all $\mu \in (\mu_0, \mu_0 + \epsilon)$ for some $\epsilon > 0$, and a supercritical Hopf bifurcation occurs at $\mu = \mu_0$.*

ii. *If at $\mu = \mu_0$ the nonlinear system is unstable around (x^*, y^*), then the system has a limit cycle for all $\mu \in (\mu_0 - \epsilon, \mu_0)$ for some $\epsilon > 0$, and a subcritical Hopf bifurcation occurs at $\mu = \mu_0$.*

Example 4.2 *Consider the predator-prey model*

$$x' = rx\left(1 - \frac{x}{k}\right) - \frac{axy}{b + x},$$
$$y' = -cy + \frac{daxy}{b + x},$$

where x is the density of prey, y the density of predators, r the per-capital growth rate of prey in the absence of predators, k is maximum carrying capacity, a the prey consumption rate, b the handling time, c the predator death rate due to other factors and d the conversion rate of food in to biomass. The term $ax/(b + x)$ is the functional response (the expected number of prey eaten by a predator per unit of time). All the parameters are assumed to positive.
The equilibrium points are solutions of the system

$$rx\left(1 - \frac{x}{k}\right) - \frac{axy}{b + x} = 0, \quad -cy + \frac{daxy}{b + x} = 0.$$

There are three equilibrium points: $(0,0)$, $(k,0)$, and $(x^*, y^*) = \left(\frac{cb}{ad-b}, \frac{bcdr(b+k)(C_0-1)}{k(ad-c)^2}\right)$, where $C_0 = \frac{adk}{c(b+k)}$. We note that (x^*, y^*) exists when $C_0 > 1$.

The Jacobian matrix evaluated at a given point (x, y) is given by

$$J = \begin{pmatrix} r\left(1 - \frac{2x}{k}\right) - \frac{aby}{(b+x)^2} & -\frac{ax}{b+x} \\ \frac{abdy}{(b+x)^2} & -c + \frac{adx}{b+x} \end{pmatrix}.$$

At the equilibrium point $(0,0)$, $J = \begin{pmatrix} r & 0 \\ 0 & -c \end{pmatrix}$ and has two eigenvalues with opposite signs, $\lambda_1 = r$ and $\lambda_2 = -c$. Then $(0,0)$ is a saddle point.

At the point $(k, 0)$, $J = \begin{pmatrix} -r & \frac{-ak}{b+k} \\ 0 & c(C_0 - 1) \end{pmatrix}$ and it is an upper triangular matrix with eigenvalues $\lambda_1 = -r$ and $\lambda_2 = c(C_0 - 1)$. When $C_0 > 1$, λ_1, and λ_2 have opposite signs and $(k, 0)$ is a saddle point. However, when $C_0 < 1$, λ_1, and λ_2 are negative and $(k, 0)$ is a stable node.

The change of stability of $(k, 0)$ as C_0 varies indicates the occurrence of a bifurcation at the point $C_0 = 1$. It is a saddle-node bifurcation.

At the point (x^*, y^*), with $C_0 > 1$, we have $J = \begin{pmatrix} \frac{cr(C_0-1)(b+k)-arbd}{ad[(C_0-1)(b+k)+b]} & -\frac{c}{d} \\ \frac{cdr(b+k)(C_0-1)}{c[(C_0-1)(b+k)+b]} & 0 \end{pmatrix}.$

Since the determinant Δ of J is positive, the stability of (x^*, y^*) depends on the sign of the trace τ of J, given by

$$\tau = \frac{cr(C_0 - 1)(b + k) - arbd}{ad[(C_0 - 1)(b + k) + b]} = \frac{c^2 r(b + k)}{adk(ad - c)}(C_0 - C_H^*),$$

with $C_H^* = 1 + \frac{abd}{c(b+k)}$. When $C_0 > C_H^*$, the trace is positive and (x^*, y^*) is unstable. However, when $C_0 < C_H^*$, $\tau < 0$ and the point (x^*, y^*) is asymptotically stable.

The change of stability of (x^*, y^*) as C_0 varies indicates the occurrence of another bifurcation in the system at the point $C_0 = C_H^*$. It is a Hopf bifurcation. In fact, the eigenvalues of J are $\lambda_{1,2} = \frac{1}{2}(\tau \pm \sqrt{\tau^2 - 4\delta})$. Since $\Delta > 0$, they are purely imaginary (i.e., cross the imaginary axis of the complex plane) when $\tau = 0$, that is, when $C_0 = C_H^*$.

4.4 Hysteresis – The Quenching Reaction

Certain dynamical systems in nature and science exhibit what is called *hysteresis* behaviour. In its simplest form, the system depends on a bifurcation parameter in such a way that the current state depends on the former history of the parameter. Typically, several equilibrium points exist for a range of parameter values.

The following case study from chemical engineering is a popular example also found in Logan, pp. 48–52 [31], from which the notation below has been adopted. The term *quenching* means rapid cooling.

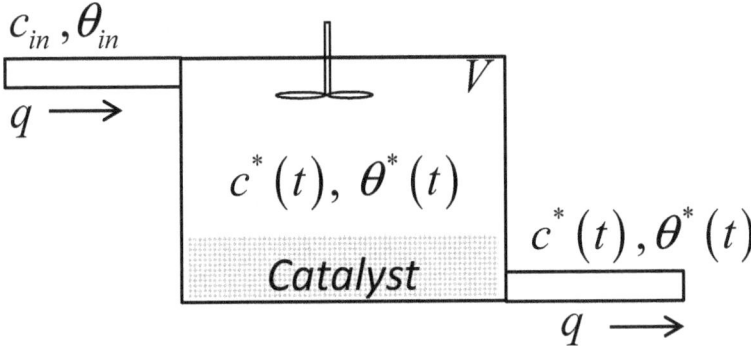

FIGURE 4.5
Sketch of well-mixed (note "whisk" symbol) and continuous flow reactor. Variables are described in the text.

A well-mixed, continuous flow chemical reactor consists of a vessel (reactor chamber) R with volume V through which there continuously flows a fluid with rate q, $[q] = \text{m}^3/\text{s}$. A chemical substance is dissolved in the fluid with entering concentration c_{in}, $[c_{in}] = \text{kg/m}^3$. Inside R there is some catalytic material causing the substance to convert into a usable product which leaves the chamber with the outflow. The reaction rate is temperature dependent, and in addition, the reaction releases heat. The contents in the chamber are vigorously stirred by means of some mechanical device, often a kind of "whisk". Well-mixed means that both temperature and concentration are constant over R. We shall establish differential equations based on conservation of substance and heat energy, and we use θ^* for the fluid's absolute temperature and refer to Figure 4.5 for the symbols. The chemical reaction depends on temperature, and for a closed vessel, the concentration of the substance changes according to

$$\frac{dc^*(t^*)}{dt^*} = -ke^{-A/\theta^*(t)}c^*(t^*).$$

This form of the temperature dependence on the reaction rate was found by the Swedish physicist Svante Arrhenius, and we shall take the equation for granted here. The reaction releases heat Q^* per volume unit proportional to the amount of converted material per volume unit, $Q^* = h(c_{in} - c^*)$. Thus,

$$\frac{dQ^*(t^*)}{dt^*} = -h\frac{dc^*(t^*)}{dt^*}.$$

Neglecting the heat associated with the substance, the heat content in the fluid may be written $Q^* = C\theta^*$ where C is the fluid's heat capacity, $[C] = \frac{\text{J}}{\text{m}^3\text{K}}$. The material and heat balance equations for the reaction chamber are now easy to state:

$$V\frac{dc^*}{dt^*} = qc_{in} - qc^* - V\left(ke^{-A/\theta^*(t)}c^*\right), \tag{4.9}$$

$$V\frac{d(C\theta^*)}{dt^*} = qC\theta_i - qC\theta^* + hV\left(ke^{-A/\theta^*(t)}c^*\right). \tag{4.10}$$

Before we proceed, it is useful to scale the equations. A reasonable choice would be to use c_i for c^* and θ_i for θ^*. For time t^*, the most obvious scale is the time it takes to fill the reactor, that is, V/q. Thus, we introduce:

$$c^* = c_{in}c,$$
$$\theta^* = \theta_{in}\theta,$$
$$t^* = \frac{V}{q}t,$$

into (4.9) and (4.10) and leave the derivation of the dimensionless equations to the reader. Then we get

$$\frac{dc}{dt} = 1 - c - \frac{c}{\mu}e^{-\gamma/\theta}, \tag{4.11}$$

$$\frac{d\theta}{dt} = 1 - \theta + \frac{bc}{\mu}e^{-\gamma/\theta}, \tag{4.12}$$

where $\mu = \frac{q}{Vk}$, $\gamma = \frac{A}{\theta_{in}}$ and $b = \frac{hc_{in}}{\theta_{in}C}$. By adding (4.11) multiplied by b and (4.12), we obtain

$$\frac{d}{dt}(bc + \theta) + (bc + \theta) = b + 1,$$

and thus,

$$bc(t) + \theta(t) = Be^{-t} + b + 1.$$

However, since we may assume that $bc(0) + \theta(0) = b + 1$, it follows that $B = 0$, and thus

$$c(t) = \frac{1 - \theta(t) + b}{b}.$$

By inserting this into (4.12) and at the same time introducing the excess temperature $u = \theta - 1$, we obtain the following equation for $u(t)$:

$$\frac{du}{dt} = -u + \frac{b - u}{\mu}e^{-\gamma/(u+1)}. \tag{4.13}$$

The equilibria are obtained as solutions of

$$\mu u = g(u) = (b - u)e^{-\gamma/(u+1)},$$

and their stability follows from the u-derivative of the right-hand side of (4.13), that is,

$$\frac{d}{du}\left(-u + \frac{g(u)}{\mu}\right) = -1 + \frac{1}{\mu}g'(u) = \begin{cases} > 0 & \text{if } g'(u) > \mu, \\ < 0 & \text{if } g'(u) < \mu. \end{cases}$$

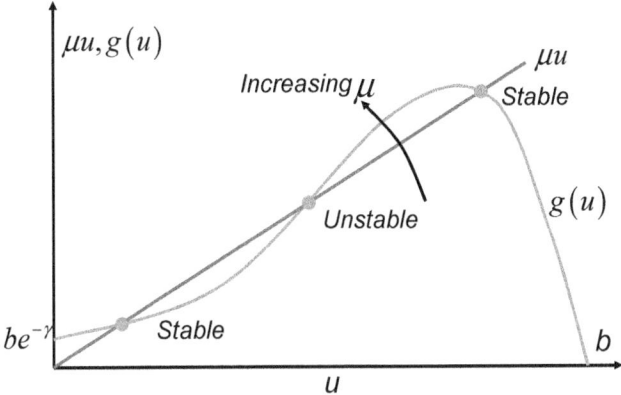

FIGURE 4.6
Solution for the equilibrium points. Stable (green) and unstable (red). The μ-parameter (slope of the blue line) acts as the bifurcation parameter.

The typical situation for a suitable set of parameters is sketched in Figure 4.6. The solutions in Figure 4.6 lead to the bifurcation diagram in Figure 4.7. The system will stay close to a stable equilibrium as long as μ changes slowly (compared to the transition times at A and B).

4.5 Exercises

1. Determine the equilibrium points and their stability for

$$\frac{\mathrm{d}u}{\mathrm{d}t} = u\left(25 - u\right).$$

2. Find the equilibrium solutions of the following differential equations and determine whether they are stable or unstable:

   (i) $\frac{\mathrm{d}u}{\mathrm{d}t} = u^2(u^2 - 1)$,
   (ii) $\frac{\mathrm{d}u}{\mathrm{d}t} = \left(u - u^2\right)\left(u - \mu\right)$, $u \ge 0, \mu \ge 0$.

3. Explain that, in order to determine the stability of an equilibrium point u_0 for the model $\mathrm{d}u/\mathrm{d}t = f\left(u\right)$, it is usually sufficient to consider the sign of $f\left(u\right)$ around u_0. Apply this when making a sketch of the bifurcation diagrams indicating stable and unstable equilibrium points in the (μ, u)-plane for the models

   (i) $\frac{\mathrm{d}u}{\mathrm{d}t} = \left(\mu + u^2 - 2u - 1\right)\left(\mu + u\right)$,
   (ii) $\frac{\mathrm{d}u}{\mathrm{d}t} = u(9 - \mu u)(\mu + 2u - u^2)$.

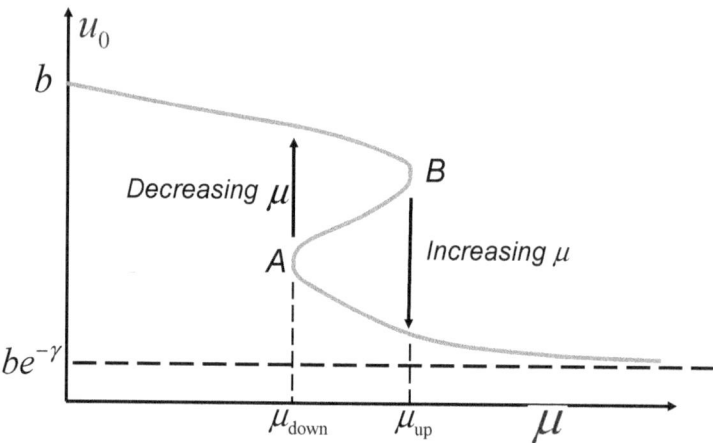

FIGURE 4.7
Bifurction diagram indicating the hysteresis behaviour

(It is not necessary to derive the exact expressions for the equilibrium points or consider points requiring higher-order analysis.)

4. A certain population develops according to the equation of logistic growth. For small populations, there is some possibility that individuals may die before they have met a partner, and this can be modelled by modifying the logistic model as follows,

$$\frac{dy}{dt} = y - y^2 - \mu y, \ \mu \geq 0. \tag{4.14}$$

(a) Determine the equilibrium solutions of Eq. 4.14 and determine whether they are stable or unstable.

(b) For larger populations, this modification will be less important. The following model has therefore been proposed:

$$\frac{dy}{dt} = y\left(1 - \frac{y}{M} - \frac{\mu}{1+y}\right).$$

Here we assume that $M > 1$ and, as above, that $\mu > 0$. Sketch the bifurcation diagram for the modified model and determine stable and unstable equilibrium solutions. Investigate what happens when the population is stable and positive, and μ grows slowly past the value $\frac{(M+1)^2}{4M}$.

5. The atmosphere of the Earth, the upper layer of the ocean (down to about 20m), and the upper few meters of the crust make up a mass M of about 3×10^{18} kg. It is this mass that continuously receives heat from the sun and loses heat by radiation into space.

The simplest model for Earth's mean temperature may therefore be written

$$\frac{Mc}{4\pi R^2}\frac{dT}{dt} = Q_0 - \sigma T^4,$$

where R is the radius of the Earth, c is the mean specific heat, Q_0 is the mean absorbed heat radiation from the sun per area unit, and σ is Stefan-Boltzmann constant. We write the equation as

$$C\frac{dT}{dt} = Q_0 - \sigma T^4,$$

and apply $Q_0/\sigma = (287\mathrm{K})^4$ and $C/\sigma = 6 \times 10^9 \mathrm{days} \times \mathrm{K}^3$ in the following

(a) Find the stationary temperature, show that it is stable, and the time (in days) it takes for some perturbation of the equilibrium temperature to die out.

(b) The amount of absorbed radiation from the sun depends on, among other things, the colour of Earth's surface (also called *albedo*). During ice ages, when a relatively larger part of Earth is covered with ice and snow, the mean absorbed energy will be less than during periods with higher temperatures. It has therefore been suggested to modify the term Q_0 to

$$Q_0 + Q_a \tanh\left(\frac{T - T_0}{T_n}\right),$$

where $Q_0 - Q_a$ represents ice ages, $Q_0 + Q_a$ warm periods, and T_0 is the equilibrium temperature from (a). The temperature T_n controls the width of the transition range. Discuss the modified model, and explain what can happen if

$$T_n < \frac{Q_a}{4Q_0}T_0$$

and Q_0 varies. *Hint*: Make a sketch and recall that

$$\tanh(x) \underset{x \to -\infty}{\longrightarrow} -1,$$
$$\tanh(x) = x + O\left(x^3\right), \text{ for small } x\text{-s},$$
$$\tanh(x) \underset{x \to \infty}{\longrightarrow} 1.$$

6. Find the general solution and sketch phase diagrams for the following systems. Characterize the type and the stability of the equilibria:

(a)

$$\dot{x} = x - 3y,$$
$$\dot{y} = -3x + y.$$

(b)

$$\dot{x} = 4y,$$
$$\dot{y} = -9x.$$

7. Suppose we want to model a two-dimensional autonomous system for $(x(t), y(t))$ with a stable steady state at $(-1, 3)$, and an unstable steady state at $(0, 5)$. Write down a suitable differential equation, and verify its properties.

8. Determine the behaviour of the solutions near the origin $(0, 0)'$ for the system

$$\dot{\mathbf{x}} = \begin{bmatrix} 3 & \beta \\ 1 & 1 \end{bmatrix} \mathbf{x},$$

   for different values of β.

9. Find the value(s) where the solution bifurcates and examine the stability of the origin of the equation

$$\dot{\mathbf{x}} = \begin{bmatrix} 1 & \mu \\ \mu & 1 \end{bmatrix} \mathbf{x}.$$

10. Consider the tritrophic dimensionless food chain model (obtained in Chapter 2, Section 1.3.7)

$$\frac{dx}{d\tau} = x(1-x) - \kappa \varphi_x(x) y,$$
$$\frac{dy}{d\tau} = \varepsilon y(1-y) + \varphi_x(x) y - \varphi_y(y) z,$$
$$\frac{dz}{d\tau} = -\gamma z + \varepsilon \varphi_y(y) z.$$

   Find the equilibrium points and study their stability, using the Holing type functional: $\phi_z(z) = \alpha z$, $\phi_z(z) = \frac{\alpha z}{b+z}$, $\phi_z(z) = \frac{\alpha z^2}{b+z^2}$, and $\phi_z(z) = \frac{\alpha z}{z^2/e+z+b}$.

5

Population Models

This chapter reviews some of the classical mathematical models for the dynamics of populations. Mechanistic in nature, population models link changes in population density and structure to reactions at the individual level. The study of population change has a very old history. In 1202, Leonardo of Pisa included an activity in one of his mathematical books that required creating a mathematical model for an increasing population of rabbits. He wondered about the number of pairs of rabbits that could be generated from a single pair in a year.

Basic population models such as exponential and logistic models are natural starting points in the study of population dynamics.

5.1 Single Species Dynamics

Let $N(t)$ be the population size at time t, $b(N)$, and $d(N)$ the birth and death rates respectively, and $f(N)$ the variation within the population (resulting from immigration, emigration, natural disaster, etc.). After a time Δt,

$$N(t + \Delta t) = N(t) + (b(N)N(t))\Delta t - (d(N)N(t))\Delta t + f(N(t))\Delta t.$$

The terms $b(N)N(t)\Delta t$ and $d(N)N(t)\Delta t$ represent the new births and new deaths in the time interval δt, respectively. Then

$$\frac{N(t + \Delta t) - N(t)}{\Delta t} = (b(N) - d(N))N(t) + f(N(t)).$$

For a small time interval Δt (i.e., when $\Delta t \to 0$), we obtain the general single species model

$$\frac{dN}{dt} = r(N)N + f(N),$$

which describes the rate of change in the population, with $r(N)$ being the growth rate.

DOI: 10.1201/9781003725206-5

5.1.1   The Malthusian growth model

Also called *simple exponential growth model*, it assumes a constant growth
rate r and no external influence ($f(N) = 0$). Named after Thomas Robert
Malthus [32], the model is formulated as

$$\frac{dN}{dt} = (b - d)N = rN, \tag{5.1}$$

where b and d are constant birth and death rates *per capita*, respectively. With
an initial population size of N_0, the solution of (5.1) is

$$N(t) = N_0 e^{rt}.$$

As a result, the population will go extinct over time if $r < 0$ (i.e., the scenario
of more deaths than births). However, when $r > 0$, the model predicts an
exponential growth, leading to the explosion of the population overtime. This
is not realistic, especially when there is limited resources. To overcome this,
Pierre Verhulst in 1838 considered the logistic.

5.1.2   The logistic equation

Verhulst [38] assumed that growth linearly depends on the population size
and is limited by the available logistic. He also considered the competition
(for space or resources) within species. He proposed the following model

$$\frac{dN}{dt} = (r - \alpha N)N = rN - \alpha N^2,$$

where α is the intraspecific competition (crowding effect) coefficient. Later,
Raymond Pearl and Lowell Reed popularized the model. They assumed that
growth at an early stage is approximately exponential (i.e., constant growth
rate), then slows with saturation and stops at maturity (when maximum capacity is reached). It is translated by

$$r(N) = r_0 - \alpha N, \quad r(0) = r_0 \text{ and } r(K) = 0,$$

where K is the carrying capacity, and implies that

$$r(N) = r_0 - \frac{r_0}{K}N = r_0\left(1 - \frac{N}{K}\right).$$

This results in the logistic equation

$$\frac{dN}{dt} = r_0 N \left(1 - \frac{N}{K}\right). \tag{5.2}$$

If $N << K$, the solution is approximately exponential with natural time scale
$1/r$. Using the separation of variables, (5.2) can be reduced to the integral
equation

$$\int \frac{K}{N(K - N)} dN = \int r\, dt,$$

which is computed using partial fraction decomposition as follows

$$\int \left(\frac{1}{N} + \frac{1}{K-N} \right) dN = \int r dt,$$

and has as solution

$$N(t) = \frac{N(0)K}{N(0) + (K - N(0))e^{-rt}}.$$

The capacity K, which is also a stationary solution, gives us a scale for N^*. This leads to the scaled equation

$$\frac{dN}{dt} = N - N^2 \tag{5.3}$$

(We use the scaling $N^* = \frac{N}{K}$ and $t^* = rt$, and we omit the superscript $*$ in the scaled equation (5.3) for notation purposes.) The most natural is to solve the equation by separation, but the simplest is to introduce $U = 1/N$, which gives the linear equation $\dot{U} + U = 1$, with general solution

$$U(t) = 1 + Ae^{-t}.$$

Thus,

$$N(t) = (1 + Ae^{-t})^{-1}.$$

Depending on the sign of A, the solution may be expressed as

$$s(t) = \frac{1}{1 + e^{-(t-t_0)}}$$

for $A > 0$, and

$$s(t) = \frac{1}{1 - e^{-(t-t_0)}}$$

for $A < 0$.

We see that *all* solutions within the region $(0,1)$ are expressible by the single function

$$s(t) = \frac{1}{1 + e^{-t}},$$

which is called the logistic curve, or the *sigmoid*. Solutions in the region above the horizontal line $u = 1$ evolve according to the single function $1/(1 - e^{-t})$, as shown in Figure 5.1. We immediately see that $N \equiv 1$ is a stable equilibrium solution, while $N \equiv 0$ is unstable. One feature of the sigmoid is that no matter how much time it has taken a population to reach a certain level, e.g., $K/10$, it will only take time $\mathcal{O}(1/r)$ to reach saturation. We also note that no matter how high the starting level is, we will reach equilibrium in time $\mathcal{O}(1/r)$. Thus, changing K significantly has dramatic consequences.

Though logistic growth models have been extensively used in population models, they have limitations as they cannot be used to understand scenarios of overcrowding or under-crowding.

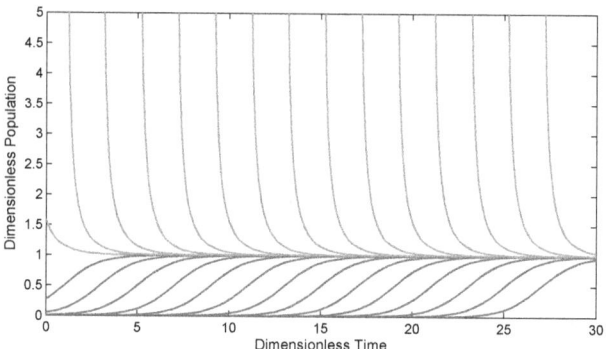

FIGURE 5.1
Solutions of the scaled and dimensionless logistic equation.

5.1.3   The Allee effect

We present in this section a model with a quadratic growth rate that can be used to describe the Allee effect and scenarios of under-crowding.

In biology, the Allee effect is a phenomenon that is defined by a relationship between population size and mean individual fitness (commonly expressed as growth rate per capita). Warder Clyde Allee, an American ecologist, was the first to describe it in the 1930s. He experimentally proved that when there are more goldfish in the tank, they grow more quickly (Allee and Bowen [1]). This led him to the conclusion that aggregation can increase an individual's chances of survival. The concept of the Allee effect challenged the classical thought of population dynamics which predicts a reduced growth rate at higher density due to competition for (limited available) resources (see logistic growth). However, within a species, individuals frequently need the support of other individuals for purposes beyond simply reproduction in order to survive. Animals that hunt together or protect against predators provide the clearest illustration of this. There is a strong and a weak Allee effect. They differ on the critical population size or density.

5.1.3.1   Strong Allee effect (with a critical population)

In this situation, the growth is negative when the population size is too large or too small (see Figure 5.2). The growth rate is expressed as follows

$$r(N) = r_0 \left(1 - \frac{N}{K}\right) \left(\frac{N}{K_1} - 1\right), \quad 0 \leq K_1 \leq K,$$

where K is the carrying capacity and K_1 is a critical size.

The strong Allee effect has been observed in the population of flour beetles (major pests in the agricultural industry and very resistant to insecticides).

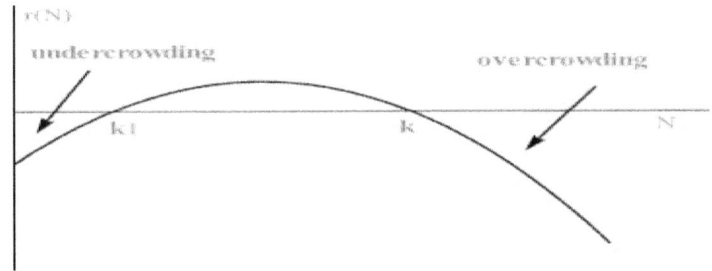

FIGURE 5.2
Growth rate curve in the case of a strong Allee effect

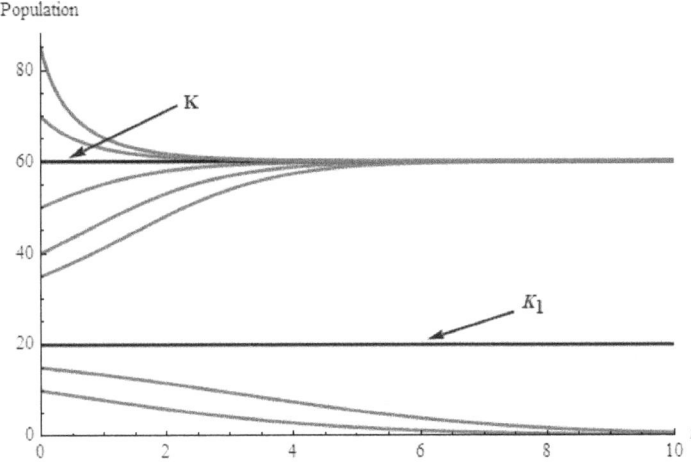

FIGURE 5.3
Time series plot of the strong Allee effect with different initial conditions.

It was found that when fewer mates are available, the female produces fewer eggs [2, 3]. The mathematical model is given by

$$\frac{dN}{dt} = r_0 N \left(1 - \frac{N}{K}\right)\left(\frac{N}{K_1} - 1\right) = f(N).$$

It has two stable equilibria at $N^* = 0$ and $N^* = K$, and one unstable at $N^* = K_1$. The solutions are illustrated in Figure 5.3. It is observed that below the critical size, the flour beetle population decreases and goes extinct over time. Above the critical size, the population increases but does not exceed the carrying capacity. However, above the capacity (overcrowding), the population decreases until saturation.

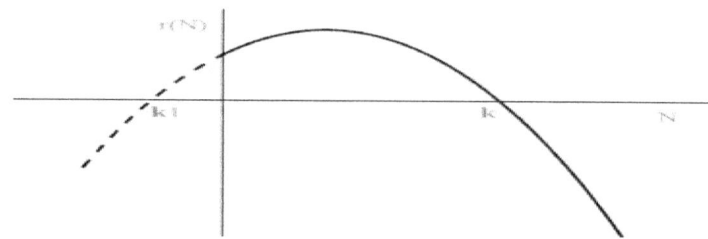

FIGURE 5.4
Growth rate curve in the case of a weak Allee effect

5.1.3.2 Weak Allee effect (no critical population)

Here, the growth is negative when the population size is too large and positive (but not at its highest) when the population size is too small (see Figure 5.4). The growth rate is expressed as follows:

$$r(N) = r_0 \left(1 - \frac{N}{K} \right) \left(\frac{N}{k_1} - 1 \right), \quad 0 \le -k_1 \le k.$$

5.1.4 Harvesting model

There are ecological, social, and economic benefits to harvesting. Socially, it might be utilized for idolatry, customary practices, and home feeding. It is economically used for profit-making in the food, clothes, and pharmaceutical sectors. Harvesting may be utilized for resource sustainability and rejuvenation from an ecological standpoint. Therefore, it's critical to establish sustainable development plans for harvesting methods; otherwise, overexploitation might exterminate certain species. It is easy to incorporate harvesting into a model, as it reduces the population size. The general model can be formulated as

$$\frac{dN}{dt} = r(N)N - h(N),$$

where $r(N)$ and $h(N)$ are the growth and harvesting rates, respectively. Gordon [19] and Schaefer [34] assumed that the catch of fish per unit effort is proportional to the stock level N. They proposed the following harvest rate function in the Gordon-Schaefer model:

$$h(N) = EN = q\xi N,$$

where E is the intrinsic harvesting rate, q the catchability coefficient, ξ the fishing effort, and N the stock level. If we consider logistic growth, the following harvesting model can be considered

$$\frac{dN}{dt} = rN \left(1 - \frac{N}{K} \right) - q\xi N.$$

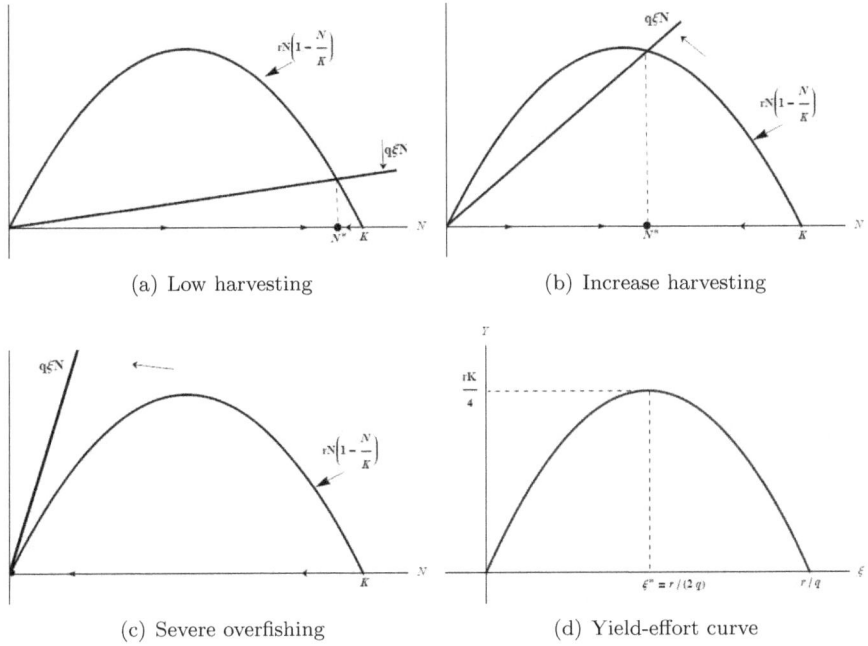

(a) Low harvesting   (b) Increase harvesting

(c) Severe overfishing   (d) Yield-effort curve

FIGURE 5.5
Phase portrait of the harvesting model

We have equilibria whenever the growth rate of the fish population balances the harvest rate, that is, when

$$rN(1 - \frac{N}{K}) = q\xi N.$$

There are two equilibria: $N^* = K(1 - \frac{q\xi}{r})$, with $q\xi < r$, and $N^* = 0$ (corresponding to the extirpation of the stock). When $q\xi < r$, the equilibrium $N^* = K(1 - \frac{q\xi}{r})$ is asymptotically stable and $N^* = 0$ is unstable. As we begin to increase the intrinsic harvesting rate $q\xi$, the equilibrium $N^* = K(1 - \frac{q\xi}{r})$ shifts to the left but maintains its stability (see Figure 5.5(b)). However, as we continue to increase $q\xi$, we eventually reach a point where the harvesting rate exceeds the growth rate for all positive stock levels. Then the origin $N^* = 0$ is the only equilibrium point, and is asymptotically stable (see Figure 5.5(c)). This happens because fish are more harvested than they grow ($q\xi > r$).

The sustainable yield at equilibrium is defined as

$$Y = q\xi N^* = q\xi K \left(1 - \frac{q\xi}{r}\right). \tag{5.4}$$

It is a quadratic function of ξ. Increasing the fishing effort increases sustainable yield - up to a point. Beyond that point, an increase in the effort lowers the yield as the stock becomes increasingly overexploited and depleted (see Figure 5.5(d)).

At the peak of the parabola in (5.4) (when $\frac{dY}{d\xi} = 0$), the maximum sustainable yield (MSY) occurs; that is, at the effort

$$\xi_{MSY} = \frac{r}{2q}.$$

This results in the maximum sustainable yield $Y_{MSY} = \frac{rK}{4}$ and the equilibrium $N^* = K/2$. Hence, optimal resource management is attained when the stock is kept at a level with maximal growth, i.e., at half the maximal stock level. This is a well-known "law" in resource management.

5.2 Multiple Species Dynamics

Understanding the dynamics of ecological systems requires an understanding of the interactions between species. Three fundamental forms of interactions exist between various species that live in the same environment: mutualism, competition, and host-parasite or predator-prey relationships. When species use the same resources and hinder one another in their quest for resources, competition arises. However, when a species' existence benefits other species, it is said to be mutualism. Within the predator-prey category, certain species (the predators) hunt and capture other species (the prey) in order to consume them. One of the first mathematical models proposed to understand these interactions and their effects on coexistence in the ecosystem is the Lotka-Volterra model.

5.2.1 Competition models

In 1934, the Russian biologist Georgyi Frantsevitch Gause (1910–1986) wrote the book *The Struggle for Existence [18]*, where he formulated *The Competitive Exclusion Principle*, which states that two species cannot coexist in the long term if they compete for the same limited resource (applicable to the current situation in the Middle East?). A simple model which reflects such a situation is the following:

$$\frac{1}{N_1^*} \frac{dN_1^*}{dt^*} = r_1 \left(1 - \alpha N_2^* \right),$$

$$\frac{1}{N_2^*} \frac{dN_2^*}{dt^*} = r_2 \left(1 - \beta N_1^* \right),$$

where N_1 and N_2 are the sizes of the two competing populations. Two obvious time scales appear, namely $1/r_1$ and $1/r_2$. If the difference between the time

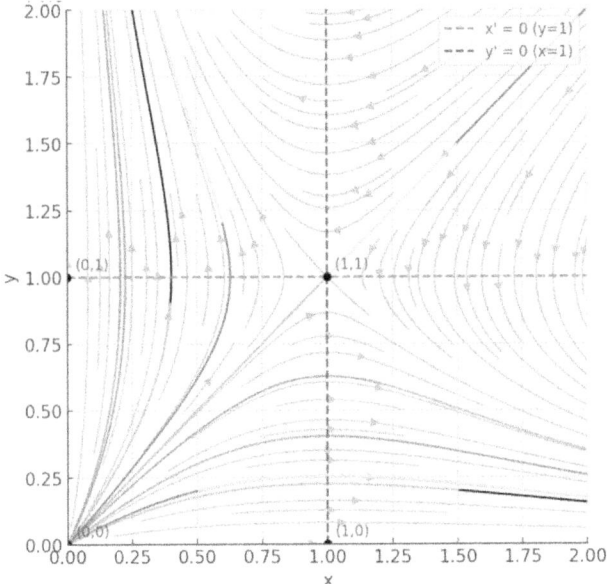

FIGURE 5.6
Phase portrait of solutions for Gause's model with $\varepsilon = 1$.

scales is large, this will be a stiff system, with behaviour characteristic of singular perturbation. The scaling is easy, and leads to a dimensionless system of the form

$$\frac{dx}{dt} = x\,(1-y),$$

$$\frac{dy}{dt} = \varepsilon y\,(1-x),$$

where the parameter ε expresses the relationship between the time scales. From the symmetry of the equations, we see that all conclusions reached about small ε can be rewritten as corresponding statements for large ε. Thus, it is enough to consider $0 < \varepsilon \leq 1$. The system has equilibrium points in (1,1) which is a saddle point, and (0,0) which is an unstable node. If one of the populations is 0, the other grows without limits. This model is not especially realistic.

A numerical solution of the system for $\varepsilon = 1$ is shown in Figure 5.6. We see that the first quadrant splits into four areas delimited by the coordinate axes and two curves crossing each other in (1,1). These curves are called *separatrices*. Systems close to the separatrix $y = x$ live dangerously: A small disturbance.

It is easy to find an implicit equation for the trajectories by dividing the equations by each other and separating the variables:

$$\left(\frac{1}{y} - 1\right) dy = \varepsilon \left(\frac{1}{x} - 1\right) dx.$$

This gives all non-trivial trajectories, expressed as

$$ye^{-y} = C\left(xe^{-x}\right)^{\varepsilon}, \ 0 < \varepsilon \leq 1,$$

where C is a positive constant. For given values of ε and C, we see from xe^{-x} that there exist four pairs of solutions. Pairs of these lie on the same trajectories. The maximum value of the left-hand side is e^{-1}. If $\max_x C(xe^{-x})^{\varepsilon} > e^{-1}$, i.e., $C > e^{\varepsilon-1}$, we describe the trajectories to the right and left of the equilibrium $(1,1)$. Otherwise, we describe the trajectories over and under $(1,1)$. The separatrices are given implicitly by

$$ye^{-y} = e^{\varepsilon-1}\left(xe^{-x}\right)^{\varepsilon}.$$

A reasonable improvement of the above model assumes that each population satisfies a logistic growth in the absence of the other. It can be formulated as follows

$$\frac{dN_1}{dt*} = r_1N_1\left(1 - \frac{N_1}{K_1} - b_{12}\frac{N_2}{K_1}\right),$$

$$\frac{dN_2}{dt*} = r_2N_2\left(1 - \frac{N_2}{K_2} - b_{21}\frac{N_1}{K_2}\right),$$

where the positive constants K_i are the capacity, r_i the intrinsic growth rate and b_{ij} the impact rate of population j on population i, $i \in \{1,2\}$. The non-dimensional system is

$$\frac{du_1}{dt} = u_1\left(1 - u_1 - \alpha_{12}u_2\right),$$

$$\frac{du_2}{dt} = \rho u_2\left(1 - u_2 - \alpha_{21}u_1\right),$$

where $\rho = \frac{r_2}{r_1}, \alpha_{12} = \frac{b_{12}K_2}{K_1}, \alpha_{21} = \frac{b_{21}r_2K_1}{r_1K_2}$. There are four equilibria:

- $(0,0)$ describing the extinction of the two populations. It is unstable.

- $(1,0)$ describing the survival of the first population and extinction of the second population. It is asymptotically stable when $\alpha_{21} > 1$ and unstable when $\alpha_{21} < 1$.

- $(0,1)$ describing the extinction of the first population and survival of the second population; It is asymptotically stable when $\alpha_{12} > 1$ and unstable when $\alpha_{12} < 1$.

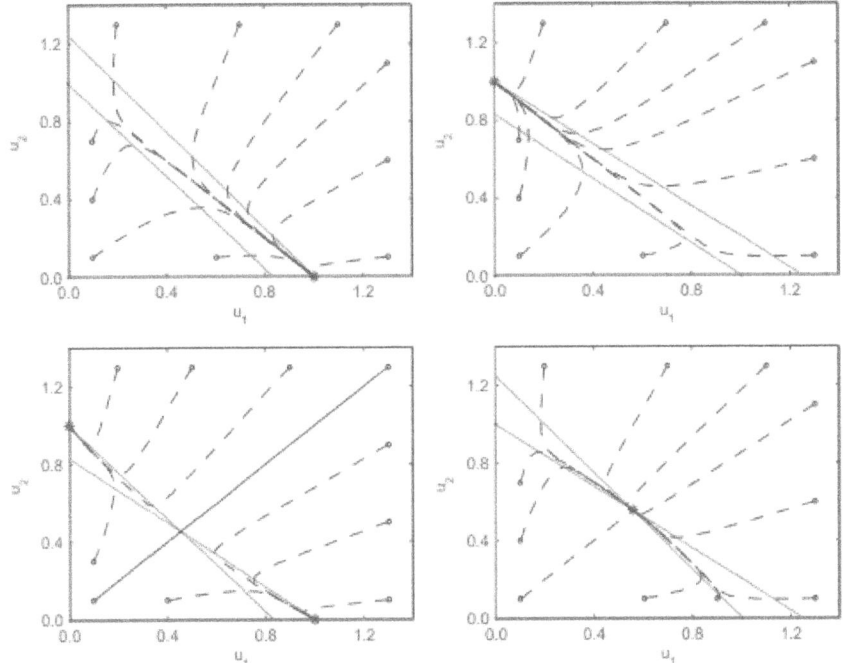

FIGURE 5.7
Dynamics of the non-dimensional competition model.

- (u_1^*, u_2^*) describing the coexistence of both populations. It is asymptotically stable when $\alpha_{12}\alpha_{21} < 1$ and unstable when $\alpha_{12}\alpha_{21} > 1$.

The phase portrait is illustrated in Figure 5.7. It is observed that on the top left (when $\alpha_{12} < 1 < \alpha_{21}$) that u_2 is excluded, and the top right (when $\alpha_{21} < 1 < \alpha_{12}$) that u_1 is excluded. Likewise, on the bottom left ($\alpha_{12} > 1$ and $\alpha_{21} > 1$), we can also have exclusion of either population depending on the initial conditions. One of the two competitors will always have a slight advantage over the other when resources are limited, leading to extinction in the long term.

5.2.2 Mutualism models

Mutualism is crucial for coexistence because it acts as a moderator in reducing the level of competition or predatory behaviour towards prey. It can increase the effectiveness of hunting. There are many examples of systems where the species have mutual benefits of each other. Plants that depend on *pollinators* is one such example. Pollinators, on the other hand, receive nectar from the plant. In mutualism scenarios, different species profit from the presence for

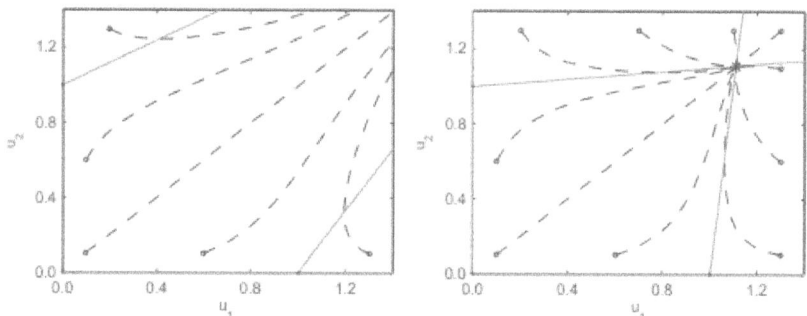

FIGURE 5.8
Dynamics of the non-dimensional mutualism system (with $\rho = 10$).

each other. A simple model is given by

$$\frac{dN_1}{dt^*} = r_1 N_1 \left(1 - \frac{N_1}{K_1} + b_{12}\frac{N_2}{K_1}\right),$$

$$\frac{dN_2}{dt^*} = r_2 N_2 \left(1 - \frac{N_2}{K_2} + b_{21}\frac{N_1}{K_2}\right),$$

leading to the non-dimensional system

$$\frac{du_1}{dt} = u_1 \left(1 - u_1 + \alpha_{12}u_2\right),$$

$$\frac{du_2}{dt} = \rho u_2 \left(1 - u_2 + \alpha_{21}u_1\right).$$

The straight line nullclines will have positive gradients leading to the two possible behaviours as shown in Figure 5.8: population explosion (on the left side, with $\alpha_{12} = 0.6 = \alpha_{21}$) and coexistence (on the right side, with $\alpha_{12} = 0.1 = \alpha_{21}$). The population explosion refers to the population that exceeds its sustainable size within a particular environment or habitat.

The following model, which is taken from [33], starts with a logistic equation for the pollinators, where the sustainable population level depends on the plant population:

$$\frac{dB^*}{dt^*} = rB^* \left(1 - \alpha\frac{B^*}{P^*}\right).$$

The plant quantity, in turn, satisfies an equation of the form

$$\frac{dP^*}{dt^*} = -qP^* + I\frac{\hat{B}}{\hat{B} + C},$$

where $\hat{B}$ is the effective density of pollinators. If there are few plants, it is not certain that the pollinators can find plants ($\hat{B} \ll B^*$), while if there are many plants, all pollinators find enough plants ($\hat{B} \approx B^*$). A possible model for $\hat{B}$ can then be

$$\hat{B} = B^* \frac{P^*}{P^* + D}.$$

Together, this gives the equations

$$\frac{dB}{dt}\frac{1}{B} = 1 - a\frac{B}{P},$$

$$\frac{dP}{dt}\frac{1}{P} = -\varepsilon + b\frac{B}{BP + P + 1},$$

where we have scaled in the following way:

$$P^* = DP, \ B^* = CB, \ a = \alpha C/D, \ b = I/(rD), \ t^* = \frac{1}{r}t, \ \varepsilon = q/r.$$

By solving for the equilibria, we end up with

$$B_0 = \frac{1}{2a\varepsilon}\left(b - \varepsilon a \pm \sqrt{(b - \varepsilon a)^2 - 4a\varepsilon^2}\right),$$

$$P_0 = aB_0.$$

There will be two equilibria in the first quadrant if $0 < b - \varepsilon a, \ 0 < (b - \varepsilon a)^2 - 4a\varepsilon^2$, or

$$0 < (I - q\alpha C)/D,$$

$$0 < \frac{(I - q\alpha C)^2 - 4\alpha q^2 CD}{D^2}.$$

Figure 5.9 shows a set of parameters giving two equilibrium points ($\varepsilon = 1/2, a = 1, b = 1.7$). If the system is left of the separatrix, it dies out. According to the reference, isolated systems of this kind are common only in areas with stable climatic conditions.

5.2.3 Predator-prey models

The Lotka-Volterra equations, also known as the *predator-prey* equations were formulated by Alfred J. Lotka and Vito Volterra independently, around 1925. Since the equations are described in detail in most books about non-linear differential equations, the presentation below is very brief. In the same way as the equations above, they can be written as a system

$$\frac{1}{N_1^*}\frac{dN_1^*}{dt^*} = r_1\left(1 - \alpha N_2^*\right),$$

$$\frac{1}{N_2^*}\frac{dN_2^*}{dt^*} = r_2\left(-1 + \beta N_1^*\right),$$

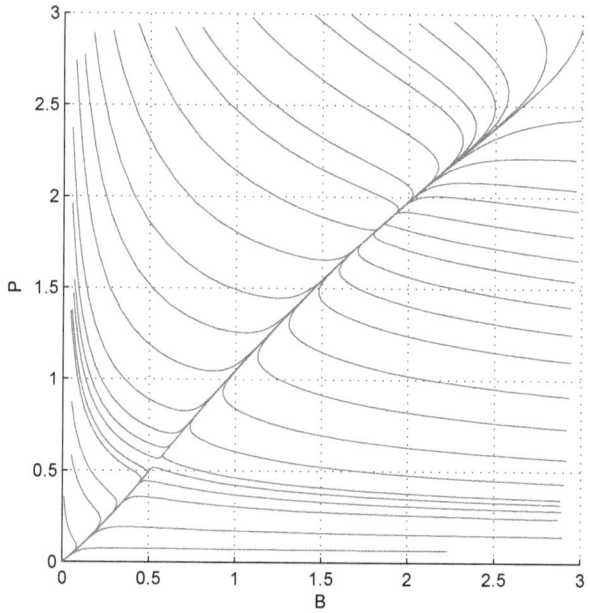

FIGURE 5.9

Phase diagram for pollinators (B) and plants (P) for a situation where we have one stable and one unstable equilibrium point.

where N_1^* is the prey density and N_2^* is the predator density. As in the preceding paragraph, after scaling, the system attains the form

$$\frac{\mathrm{d}x}{\mathrm{d}t} = x\left(1 - y\right),$$

$$\frac{\mathrm{d}y}{\mathrm{d}t} = \varepsilon y\left(-1 + x\right),$$

with equilibria in (1,1) and (0,0) independent of the size of ε. The first one is a centre, and the second a saddle point. The trajectories are shown in Figure 5.10.

The implicit equation for the trajectories is

$$ye^{-y}\left(xe^{-x}\right)^{\varepsilon} = C, \quad 0 < C.$$

Since the left-hand side is limited from above by $e^{-1-\varepsilon}$, we must have $C < e^{-1-\varepsilon}$. For a given possible value of C, x will also be restricted to the interval around $x = 1$ which satisfies $\left(xe^{-x}\right)^{\varepsilon} \geq Ce$. We get a similar interval for y defined by $ye^{-y} \geq Ce^{\varepsilon}$. Thus, the trajectory is inside a rectangle which is defined by the equality of the respective inequalities. The shape of the rectangle is determined by ε and C. When C approaches its maximum value,

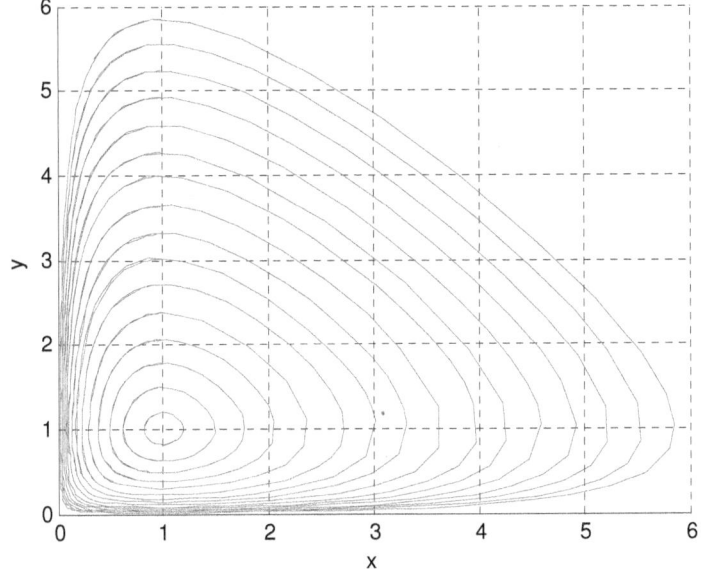

FIGURE 5.10
Paths of the Lotka–Volterra system when $\varepsilon = 1$ calculated numerically. The numerical solution goes around the orbits several times and does not connect perfectly.

the trajectories shrink towards $(1, 1)$ and become approximately elliptical with centre $(1, 1)$. When $C \ll 1$ the trajectory shifts towards the coordinate axes, unless $y \approx -\varepsilon x - \ln(C)$, which is the trajectory far away from the coordinate axes. The orbit shape is roughly triangular.

Limitations in growth, as in the logistic equation, lead to scaled equations of the form

$$\frac{\mathrm{d}x}{\mathrm{d}t} = x\left(1 - y - \alpha x\right),$$

$$\frac{\mathrm{d}y}{\mathrm{d}t} = \varepsilon y\left(-1 + x - \beta y\right),$$

with equilibria in the origin and (x_0, y_0) where

$$x_0 = \frac{1 + \beta}{1 + \alpha\beta}, \quad y_0 = \frac{1 - \alpha}{1 + \alpha\beta}.$$

We leave it to the reader to show that (x_0, y_0) becomes a stable focus when α and β are small. What else happens?

If we return to the original Lotka–Volterra equations and utilize the fact that the trajectories are periodic with period T, we get

$$\frac{1}{r_1} \int_{t^*=0}^{T} \frac{\mathrm{d}N_1^*}{N_1^*} = \int_{t^*=0}^{T} \left(1 - \alpha N_2^*\right) \mathrm{d}t^* = T - \alpha \int_{t^*=0}^{T} N_2^* \, \mathrm{d}t^*,$$

and correspondingly for the other equation. But, since

$$\frac{1}{r_1} \int_{t^*=0}^{T} \frac{\mathrm{d}N_1^*}{N_1^*} = \frac{1}{r_1} \left(\ln(N_1^*(T)) - \ln(N_1^*(0)) \right) = 0,$$

so

$$\frac{1}{T} \int_{t^*=0}^{T} N_2^* \, \mathrm{d}t^* = \frac{1}{\alpha},$$

$$\frac{1}{T} \int_{t^*=0}^{T} N_1^* \, \mathrm{d}t^* = \frac{1}{\beta}.$$

The average levels become equal to the values at the equilibrium point.

If one studies the predator-prey models somewhat more in depth, one will see that the Lotka–Volterra equations are special. Typically, such models will have stable equilibrium points. If one observes oscillations in nature, these will not be periodic solutions about a neutral equilibrium point, but rather so-called *stable limit cycles* which are more stable to perturbations.

Delays in the Lotka–Volterra models have also been studied, and unlike ordinary 2-dimensional ODE systems, such equations can actually have *chaotic* behaviour.

It is also possible to study what hunting means to a Lotka–Volterra system. If we assume constant capture relative to population, this can be modelled as

$$\frac{\mathrm{d}N_1^*}{\mathrm{d}t^*} \frac{1}{N_1^*} = r_1 \left(1 - \alpha N_2^*\right) - f_1,$$

$$\frac{\mathrm{d}N_2^*}{\mathrm{d}t^*} \frac{1}{N_2^*} = r_2 \left(-1 + \beta N_1^*\right) - f_2.$$

Since we can write

$$\frac{\mathrm{d}N_1^*}{\mathrm{d}t^*} \frac{1}{N_1^*} = (r_1 - f_1) \left(1 - \frac{r_1 \alpha}{r_1 - f_1} N_2^*\right),$$

$$\frac{\mathrm{d}N_2^*}{\mathrm{d}t^*} \frac{1}{N_2^*} = (r_2 + f_2) \left(-1 + \frac{r_2 \beta}{r_2 + f_2} N_1^*\right),$$

we see that as long as the parameters are constant, the behaviour will be like for a Lotka–Volterra system with modified parameters. In particular,

$$\left\langle N_2^* \right\rangle = \frac{1}{T} \int_{t^*=0}^{T} N_2^* \, dt^* = \frac{1}{\alpha} \frac{r_1 - f_1}{r_1},$$

$$\left\langle N_1^* \right\rangle = \frac{1}{T} \int_{t^*=0}^{T} N_1^* \, dt^* = \frac{1}{\beta} \frac{r_2 + f_2}{r_2}.$$

The model is not necessarily realistic if one looks at the average catch per time unit:

$$F_1 = \left\langle N_1^* \right\rangle f_1 = \frac{1}{\beta} \frac{r_2 + f_2}{r_2} f_1,$$

$$F_2 = \left\langle N_2^* \right\rangle f_2 = \frac{1}{\alpha} \frac{r_1 - f_1}{r_1} f_2.$$

We observe that if we just catch predators ($f_1 = 0$), we will be able to capture an unlimited amount, without the average level changing. In contrast, the average of the prey population will grow (!).

There is much more that could be studied for such equations, *e.g.*, the behaviour for time-dependent catch with variations that are long and short relative to the period of the stock variations.

5.3   Whales and Krill

In 1979, R. M. May [33] formulated a model for the whale-krill system in Antarctica, where N^* is the krill population and H^* the whale population:

$$\frac{dN^*}{dt^*} \frac{1}{N^*} = r \left(1 - \frac{N^*}{K_N} \right) - a_2 H^* - u_N F_N,$$

$$\frac{dH^*}{dt^*} \frac{1}{H^*} = q \left(1 - \frac{H^*}{\alpha N^*} \right) - u_H F_H.$$

As seen, the maximum sustainable level of the whale stock is proportional to the krill level. The growth rates r and q are clearly quite different. Thus, we expect $1/r \ll 1/q$, and $\varepsilon = q/r \ll 1$. If we scale based on the time scale for changes in the whale population, we end up with the following singularly perturbed system:

$$\varepsilon \dot{N} = N \left(1 - N - \gamma H - f_N \right),$$

$$\dot{H} = H \left(1 - H/N - f_H \right).$$

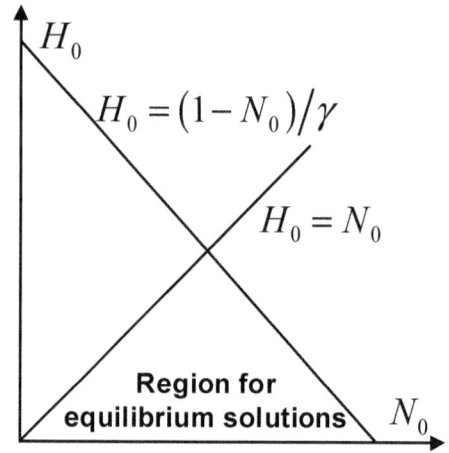

FIGURE 5.11
All equilibrium solutions lie within the shaded area when f_N and f_H are between 0 and 1.

We assume that f_N and f_H are constants between 0 and 1. Then the model has a stationary point in

$$N_0 = \frac{1 - f_N}{1 + \gamma(1 - f_H)},$$

$$H_0 = \frac{(1 - f_N)(1 - f_H)}{1 + \gamma(1 - f_H)},$$

and all singular points lie inside a triangle, as pointed out in Figure 5.11.

The equilibrium solutions are stable focuses, as shown in Figure 5.12. When ε is small, the solutions bear the characteristic sign of *singular perturbation*. This behaviour is not so striking in areas where $N = O(\varepsilon)$ or $N \ll H$, since one then also will have other small terms in the equations. If the population of whales is lower than the equilibrium level, one sees how the system quickly approaches a "quasi-static equilibrium", approximately given by $1 - N - \gamma H - f_N = 0$, and then follows it towards the equilibrium point. Obviously, this is due to the krill population reacting quickly compared to the whale stock.

For a given catch rate, the amount caught per time unit is

$$P_N = f_N N_0 = \frac{f_N(1 - f_N)}{1 + \gamma(1 - f_H)},$$

$$P_H = f_H H_0 = \frac{f_H(1 - f_H)(1 - f_N)}{1 + \gamma(1 - f_H)}.$$

We see that the maximum amount of krill we can take out is given by $f_N = 1/2$ whatever the catch of whales, while the maximum amount of whales

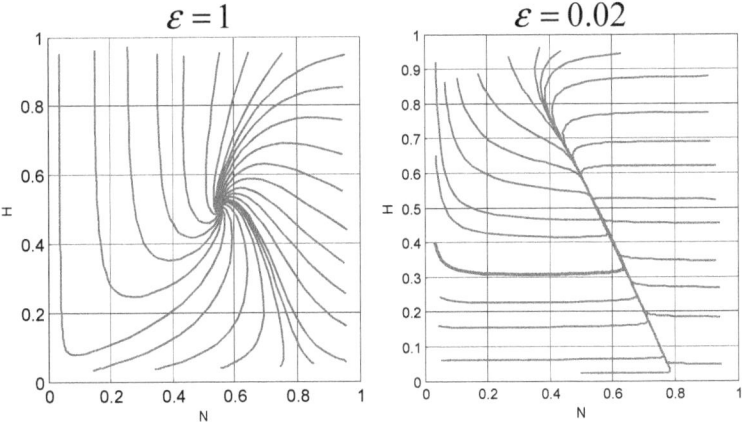

FIGURE 5.12
Phase plot of the whale-krill system with $f_N = 0.2$ and $f_H = 0.1$, $\gamma = 0.5$ with $\varepsilon = 1$ to the left, $\varepsilon = 0.02$ to the right.

is $f_H = (1 + \gamma\sqrt{1+\gamma})/\gamma$. It is not possible to increase whaling by fishing krill, while it of course is beneficial to catch whales to be able to catch more krill.

It is easy to imagine economic issues related to such a model. If one catches both whales and krill, one is interested in maximizing profits. If the prices are p_N and p_H, one would like to maximize economic return per time unit, $p_N P_N + p_H P_H$. At the same time there's a limit to the weight of catch over a period T that can be transported with available tonnage, $(w_N P_N + w_H P_H)t \leq L$.

If catch rates are proportional to the inverse of the populations, i.e., a fixed quantity is fished per unit of time irrespective of the size of the stocks, the equations are more cumbersome. Any equilibrium solutions can be found from

$$N(1 - N - \gamma H - a_N/N) = N - NH^2 - \gamma HN - a_N = 0,$$
$$H(1 - H/N - a_H/H) = H - H^2/N - a_H = 0.$$

If we assume that $\gamma = 1$ and that we only catch whales, we deduce that

$$N_0 = 1 - H_0,$$
$$H_0 = \frac{1 + a_H \pm \sqrt{(1 + a_H)^2 - 8a_H}}{4}.$$

In other words, if $0 < a_H < 3 - 2\sqrt{2}$, we have two equilibrium points. We leave details to the reader.

An important aspect of such models is whether they are stable against disturbances, and how well the catch can be controlled by decree, etc. As one understands, there is unlimited potential for mathematical, numerical, and economic analysis.

5.4 Exercises

1. Let $0 < \alpha < r$, $N_0 > 0$, and consider the following equation:

$$\frac{\mathrm{d}N^*(t^*)}{\mathrm{d}t^*} = rN^*(t^*)\left(1 - \frac{N^*(t^*)}{K}\right) - \alpha N^*(t^*), \quad t^* > 0, \ N^*(0) = N_0.$$

$$(5.5)$$

(a) What could the problem in (5.5) model? Find reasonable scales for this problem when $N_0 \gg K$ and $\alpha < r$.

(b) Let $\varepsilon, \alpha, \beta > 0$, $\alpha, \beta < 1$, and consider the following scaled system of equations for two interacting populations x and y:

$$\frac{\mathrm{d}x}{\mathrm{d}t} = x(1 - x) - \alpha xy,$$

$$\varepsilon \frac{\mathrm{d}y}{\mathrm{d}t} = y(1 - y) - \beta xy.$$

Discuss the properties of the model in terms of the parameters ε, α, and β. Find the equilibrium points and determine their stability when $\beta = 0$.

2. The following scaled model has been proposed for the population of females (F) and males (M) in a fish population:

$$\dot{F} = -F + \gamma\left(1 - e^{-M}\right)F,$$
$$\dot{M} = -M + \gamma\left(1 - e^{-M}\right)F,$$

where γ is a positive constant.

(a) Explain some of the properties of the model. For example

   i. What happens if F or M is 0?

   ii. Is there a difference in the birth rate for females and males?

   iii. How are the equations when M is small and when M is large, and how are the conditions reflecting this?

(b) What kind of point is the stationary point $(0,0)$?

(c) Show that, if $\gamma > 1$, there is another stationary point (F_0, M_0) where $F_0 = M_0$.

(d) Identify the type of the point in **(c)**.

(e) Is the model really reasonable?

3. The Eurasian elk or moose (*Alces alces*) is the largest species in the deer family. Each year over 8000 moose calves and adults are caught during the hunting season in the forests around Trondheim. Consider the following simple model for the moose population, P:

$$\frac{dP}{dt} = kP\left(1 - \frac{P}{M}\right)\left(\frac{P}{m} - 1\right), \quad 0 < m < M. \qquad (5.6)$$

(a) What properties does the model try to explain, and what are the equilibrium populations? Determine, using linear stability analysis, whether they are stable or unstable. Make a simple sketch showing the solutions of (5.6).

(b) A simplified model, also including the hunter population (J) has, after scaling, the form

$$\frac{dP}{dt} = P(1 - P) - J,$$

$$\frac{dJ}{dt} = -\frac{J}{2} + JP. \qquad (5.7)$$

Determine the equilibrium points for (5.7) and of what kind they are (or appear to be).

(c) Show that the solution of

$$h(P, J) = J - 3P(1 - P)/2 = 0$$

defines a possible path for the model in (5.7). (*Hint*: Show that ∇h along this path is always orthogonal to the direction of the motion defined by $(dP/dt, dJ/dt)$.)

(d) The point $(P_c, J_c) = \left(\frac{1}{2}, \frac{1}{4}\right)$ for (5.7) could be a centre. Shift the origin by introducing

$$x = P - 1/2,$$
$$y = J - 1/4,$$

and show that the system has paths which are symmetric about the new y-axis, and therefore that the point *is* really a centre.

(e) Summarize the qualitative properties of the model in (5.7).

4. In a population consisting of M_0 individuals, an influenza infection is transferred by sick persons ($S_0^*(t)$ individuals) meeting susceptible persons ($M_0 - S_0^*(t)$ individuals). The number of people infected per time unit is proportional to the probability that sick persons meet susceptible persons. The proportionality constant, r is called the *infection rate* (per individual). After a while, the sick individuals recover.

(a) The following dynamic model has been proposed for the number of sick persons in the population if one assumes that none are immune:

$$\frac{\mathrm{d}S^*}{\mathrm{d}t^*} = rS^* (M_0 - S^*) - \alpha S^*.$$

Consider a simplified equation in the early stages of the epidemic, and find suitable scales for the variables. The infection rate can be controlled by vaccines and other restrictions by the health authorities. Discuss the stationary solutions of the scaled equation in the light of the size of r.

(b) A more realistic model takes into account that those who have been cured will be immune, at least for some time after the illness. The following (scaled) mathematical model has been proposed for the number of sick, $S(t)$, and immune, $I(t)$, persons:

$$\frac{\mathrm{d}S}{\mathrm{d}t} = S(1 - I - S) - \lambda S,$$

$$\frac{\mathrm{d}I}{\mathrm{d}t} = \lambda S - \mu I. \tag{5.8}$$

State the region in the (S, I)-plane for physically acceptable solutions, and the meaning of the parameters μ and λ. This dynamical system has an obvious stationary point. When is this point stable? It is not necessary to study limit cases.

(c) Determine for which other values of λ and μ the system has another, physically acceptable stationary point, (S_0, I_0). Show that this point moves on a line segment if λ is kept fixed while μ varies.

(d) Linearize the system in (5.8) around (S_0, I_0) by introducing $S(t) = S_0 + x(t)$, $I(t) = I_0 + y(t)$ and show that the matrix $\mathbf{A}$ in the linearized system $\dot{\mathbf{x}} = \mathbf{A}\mathbf{x}$, $\mathbf{x} = [x, y]'$, has the form

$$\mathbf{A} = \begin{bmatrix} -S_0 & -S_0 \\ \lambda & -\mu \end{bmatrix}.$$

Determine whether (S_0, I_0) is stable or unstable.

6

Epidemiological Models

6.1   Introduction

Epidemiological models are used to describe the spread of infectious diseases in populations. They trace their origin back to Daniel Bernoulli [8], who in 1760 published a paper on immunization against smallpox. In this chapter, we discuss two well-known models: the Kermack-McKendrick model and the SIR model.

6.2   The Kermack-McKendrick Model

This model appeared in a particular case in a paper by Kermack and McKendrick in 1927 [26]. Similar to the multiple species dynamics models discussed in section 5.2, the Kermack-McKendrick model is classified as a compartmental model. This means that the population is divided into compartments, where each compartment represents a group of individuals who share common characteristics. The flow of individuals between compartments is described by a system of ordinary differential equations. To formulate the model, we begin by defining the compartments and subsequently describe the flows between them.

This model addresses the spread of an infectious disease within a population. Accordingly, the compartments considered are as follows:

- Individuals susceptible to the disease, who have yet to become infected;

- Individuals infected and infectious with the disease;

- Individuals who are removed from the infectious compartment, either due to recovery from the disease or as a result of death caused by it.

Upon infection, an individual in the compartment of susceptible individuals (which we call the susceptible compartment from now on for convenience) moves to the compartment of infected individuals. After some time in the

DOI: 10.1201/9781003725206-6

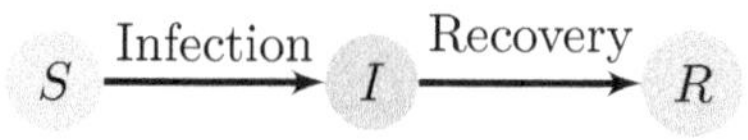

$$S \xrightarrow{\text{Infection}} I \xrightarrow{\text{Recovery}} R$$

FIGURE 6.1
Conceptual flow diagram of the Kermack-McKendrick model.

infected compartment, the individual moves to the removed compartment. This is summarized in Figure 6.1.

To translate the problem into a mathematical model, we define state variables $S(t)$, $I(t)$, and $R(t)$ as the numbers of individuals in the susceptible, infected, and removed compartments at time $t \geq 0$, respectively. The dependence on t is dropped if it is not explicitly required, and we write S, I, R. To obtain a system of differential equations, the flows between compartments are described using the following assumptions:

- The period of time under consideration is sufficiently short that demography can be neglected; we also say the model has *no vital dynamics*. This means that there is no demographical component to the change in the number of individuals in the population.

- The rate at which individuals become infected is described by a mass action incidence function; see below.

- The rate at which individuals recover from the disease is the *per capita* rate γ.

Let us return briefly to the rate at which new infections occur. In (medical) epidemiology, *incidence* is defined as the number of new cases per time unit. So in the mathematical model, the incidence function describes the rate of change of the number of infected individuals. If there are S susceptible individuals and I infectious individuals in the population, we use a function of the form

$$\beta SI.$$

This is called a mass action incidence function because it is proportional to the product of the number of susceptible and infectious individuals. There are other types of incidence functions; see Exercises.

In view of these assumptions, the flow diagram of the Kermack-McKendrick model is refined as in Figure 6.2. The resulting model is called

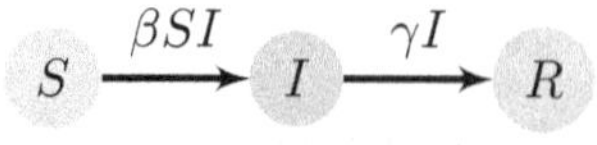

$$S \xrightarrow{\beta SI} I \xrightarrow{\gamma I} R$$

FIGURE 6.2
Flow diagram of the Kermack-McKendrick model (6.1).

the Kermack-McKendrick (KMK) SIR model. Its dynamics is governed by the following system of three ordinary differential equations:

$$S' = -\beta SI, \tag{6.1a}$$
$$I' = \beta SI - \gamma I, \tag{6.1b}$$
$$R' = \gamma I. \tag{6.1c}$$

6.2.1 Mathematical analysis

Model (6.1) has 3 compartments, but we notice that *removed* does not have a direct influence on the dynamics of S or I: R does not appear in (6.1a) or (6.1b). Furthermore, the total population (including deceased who are also in R) $N = S + I + R$ satisfies

$$N' = (S + I + R)' = 0.$$

Thus, N is constant and

$$S(t) + I(t) + R(t) = N_0, \quad t \geq 0,$$

so the dynamics of R can be deduced from $R = N - (S + I)$. As a result, we can consider the reduced system

$$S' = -\beta SI, \tag{6.2a}$$
$$I' = \beta SI - \gamma I. \tag{6.2b}$$

Let us consider equilibria of (6.2). From (6.2b), either $S^\star = \gamma/\beta$ or $I^\star = 0$. Substituting into (6.2a), we get, respectively, $(S^\star, I^\star) = (\gamma/\beta, 0)$ or that any $S^\star \geq 0$ is an equilibrium point.

The second case is an *issue*: the usual linearization does not work when there is a *continuum* of equilibria as they are not *isolated*.

Proposition 6.1 *The Kermack-McKendrick model (6.1) has the continuum of equilibria*

$$E_0^{KMK} := \{(S^\star, I^\star, R^\star) = (S_\infty, 0, N_0 - S_\infty), \quad S_\infty \in [0, N_0]\}. \tag{6.3}$$

Proof *Let us consider (6.1) and start with $I = I^\star = 0$. Substitute this value into (6.1a) at equilibrium, giving $0 = -\gamma S^\star I^\star (= 0)$, meaning that any value of $S^\star$ satisfies this relation. From the conservation of the total population (6.2), the equilibrium E_0^{KMK} takes the form given by (6.3).*

Now consider $S = S^\star = \gamma/\beta$. Substituting this value into (6.1a) at equilibrium gives $0 = -\gamma I^\star$, from which it follows that $I^\star = 0$, and, using the conservation of total population (6.2),

$$(S^\star, I^\star, R^\star) = \left(\frac{\gamma}{\beta}, 0, N_0 - \frac{\gamma}{\beta}\right) \tag{6.4}$$

is an equilibrium of (6.1). *The equilibrium* (6.4) *is biologically relevant only when* $N_0 - \gamma/\beta \geq 0$. *Note that* (6.3) *includes* (6.4) *when the latter is biologically relevant.*

The reason why the presence of a continuum of equilibria is an issue can be understood by considering the definition of the stability of equilibria. Adapting slightly the definitions in [24], consider the ordinary differential equation

$$\frac{dx}{dt} = f(x),\tag{6.5}$$

where $x(t) \in W$ and $f : W \to E$ is a function such that solutions to (6.5) exist uniquely, e.g., a C^1 function, from an open set W of the vector space E into E. Denote $x(t, x_0)$ the solution to (6.5) through the initial value $x(t_0) = x_0$. As we have already seen, a point $x^\star \in W$ is an equilibrium if $f(x^\star) = 0$. Stability of the equilibrium is then defined as follows.

Definition 6.1 (Locally stable equilibrium) *An equilibrium point $x^\star$ of* (6.5) *is locally stable if for every neighbourhood $\mathcal{N}(x^\star)$ of $x^\star$ in W, there is a neighbourhood $\mathcal{N}_1 \subseteq \mathcal{N}(x^\star)$ of $x^\star$ such that every solution $x(t, x_0)$ with $x_0 \in \mathcal{N}_1$ is defined and in $\mathcal{N}(x^\star)$ for all $t > t_0$.*

Definition 6.2 (Locally asymptotically stable equilibrium) *If $\mathcal{N}_1$ can be chosen so that in addition to the properties in Definition 6.1, $\lim_{t \to \infty} x(t, x_0) = x^\star$ for all $x_0 \in \mathcal{N}_1$, then $x^\star$ is locally asymptotically stable.*

The disease-free equilibria (6.3) of (6.1) are not isolated: any (open) neighbourhood of an equilibrium contains infinitely many other equilibria.

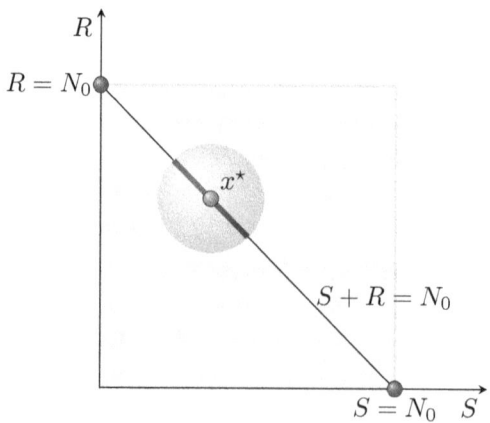

FIGURE 6.3
Neighbourhood $\mathcal{N}(x^\star)$ of $x^\star \in E_0^{\mathrm{KMK}}$ lying on the $S - R$ plane (the neighbourhood extends above and below the $S - R$ plane in the I direction, not shown here). The thin line is E_0^{KMK}, the thick line is $E_0^{\mathrm{KMK}} \cap \mathcal{N}(x^\star)$.

Proposition 6.2 *Consider a disease-free equilibrium $x^\star \in E_0^{KMK}$ of (6.1). Then $x^\star$ is locally stable but not locally asymptotically stable.*

Proof *Let $x_1^\star \in E_0^{KMK}$ be an equilibrium of (6.1). Consider $\mathcal{S}_N(x_1^\star) \subset E_0^{KMK}$, open subset of E_0^{KMK} containing $x_1^\star$. Now take some $x_2^\star \in \mathcal{S}_N(x_1^\star)$. Since $x_2^\star \in \mathcal{S}_N(x_1^\star) \subset E_0^{KMK}$, $x_2^\star$ is an equilibrium of (6.1) and thus $x(t, x_2^\star) = x_2^\star \in \mathcal{S}_N(x_1^\star)$ for all $t \geq t_0$. As a consequence, $x_1^\star$ is locally stable.*

This implies that any open neighbourhood $\mathcal{N}(x_1^\star)$ contains $\mathcal{S}_N = \mathcal{N}(x_1^\star) \cap E_0^{KMK}$. Consider, then, some $x_2^\star \in \mathcal{S}_N$. Since $x_2^\star \in \mathcal{S}_N$, $x_2^\star$ is an equilibrium and as a consequence, $\lim_{t\to\infty} x(t, x_2^\star) = x_2^\star$. Therefore, any open neighbourhood of $x_1^\star$ contains points x_0 such that $\lim_{t\to\infty} x(t, x_0) = x_1^\star$; implying that $x_1^\star$ is locally stable but not locally asymptotically stable.

This means in particular that considering the Jacobian of (6.1) at the disease-free equilibrium **makes no sense!**

6.2.2 A workaround – study I'/S'

Kermack and McKendrick remark in their 1927 paper [26] that the dynamics of dI/dS can be studied to understand the behaviour of the system. We have

$$\frac{dI}{dS} = \frac{dI}{dt}\frac{dt}{dS} = \frac{I'}{S'} = \frac{\beta SI - \gamma I}{-\beta SI} = \frac{\gamma}{\beta S} - 1, \tag{6.6}$$

provided $S \neq 0$. Recall that S and I are $S(t)$ and $I(t)$. Therefore, (6.6) describes the relation between S and I over solutions to the original ODE (6.2).

Integrating (6.6) gives trajectories in state space,

$$I(S) = \frac{\gamma}{\beta} \ln S - S + C,$$

with $C \in \mathbb{R}$. This is the equation of a family of curves in the $S - I$ plane, each corresponding to a different value of C. The curves are hyperbolas with asymptotes $I = \frac{\gamma}{\beta} \ln S$ and $I = S + C$. The equilibrium $x^\star$ is the point where the hyperbolas intersect the line $I = S$. Using the initial conditions S_0 and I_0, we can determine the value of C and thus the trajectory in the $S - I$ plane: since $I(S_0) = I_0$, it follows that $C = S_0 + I_0 - (\gamma \ln S_0)/\beta$ and the solution to (6.1) is, as a function of S,

$$I(S) = S_0 + I_0 - S + \frac{\gamma}{\beta} \ln \frac{S}{S_0},$$

$$R(S) = N - S - I(S) = R_0 - \frac{\gamma}{\beta} \ln \frac{S}{S_0},$$

since $N_0 = S_0 + I_0 + R_0$.

Let us study

$$I(S) = S_0 + I_0 - S + \frac{\gamma}{\beta} \ln \frac{S}{S_0}.$$

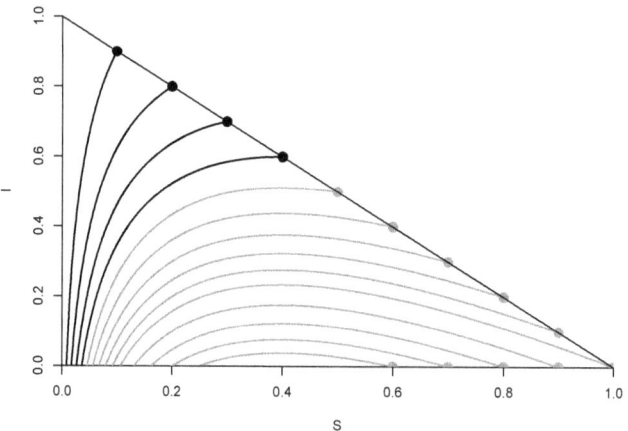

FIGURE 6.4
Trajectories of (6.2) in (S, I)-space, normalized, with initial conditions $(S_0, 1 - S_0)$ shown as dots and $\beta/\gamma = 2.5$. Red curves have $\mathcal{R}_0 > 1$ and see an epidemic, black curves have $\mathcal{R}_0 > 1$ and do not experience an epidemic.

Recall that we have

$$\frac{d}{dS}I(S) = \frac{\gamma}{\beta S} - 1.$$

So, in the previous curves, the maximum $I_{\max}$ of $I(S)$ happens when $S = \gamma/\beta$. At that point,

$$I_{\max} = I_0 + \left(1 - \frac{1}{\mathcal{R}_0} - \frac{\ln(\mathcal{R}_0)}{\mathcal{R}_0}\right)S_0.$$

Theorem 6.1 (Epidemic or no epidemic?) *Let* $(S(t), I(t))$ *be a solution to (6.2) and* $\mathcal{R}_0$ *defined by*

$$\mathcal{R}_0 = \frac{\beta}{\gamma}S_0. \tag{6.7}$$

Then

- *if* $\mathcal{R}_0 \leq 1$, *then* $I(t)$ *decreases to* 0 *when* $t \to \infty$,

- *if* $\mathcal{R}_0 > 1$, *then* $I(t)$ *first reaches a maximum* $I_{\max}$ *then goes to 0 as* $t \to \infty$.

6.2.3   The basic reproduction number $\mathcal{R}_0$

The previous discussion has highlighted the role played by the quantity $\mathcal{R}_0$ in determining the outcome of an epidemic. This indicator is often used in

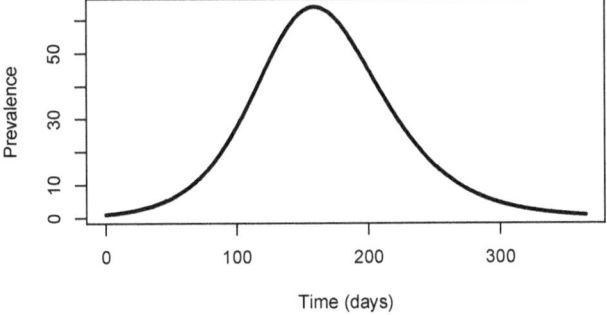

FIGURE 6.5
One solution to the Kermack-McKendrick model, when $\mathcal{R}_0 = 1.5$.

TABLE 6.1
Values of $\mathcal{R}_0$ for measles in different locations and periods (from [5]).

Infection	Location	Period	$\mathcal{R}_0$
Measles	Cirencester, England	1947-50	13-14
	England and Wales	1950-68	16-18
	Kansas, USA	1918-21	5-6
	Ontario, Canada	1912-3	11-12
	Willesden, England	1912-3	11-12
	Ghana	1960-8	14-15
	East Nigeria	1960-8	16-17

epidemiology. Verbally, $\mathcal{R}_0$ is the average number of secondary cases of infection produced when a single infectious individual is introduced into a wholly susceptible population.

$\mathcal{R}_0$ can be estimated from data. Table 6.1 lists several values for measles (a highly infectious disease) given in [5].

6.2.4 The final size of an epidemic

The final size of an epidemic is the number of individuals that will have been infected by the end of the epidemic. Again, this is considered as early as the paper of Kermack and McKendrick. This is an important question, as it allows us to estimate the impact of an epidemic on a population.

In preparation for the remainder of this section, for a nonnegative valued integrable function $w(t)$, denote

$$w_0 = w(0), \qquad w_\infty = \lim_{t \to \infty} w(t), \qquad \hat{w} = \int_0^\infty w(t) \, dt.$$

Returning to subsystem (6.2), compute the sum of (6.2a) and (6.2b), making sure to show time dependence:

$$\frac{d}{dt}(S(t) + I(t)) = -\gamma I(t).$$

Now integrate both sides of this expression from 0 to ∞, giving

$$\int_0^\infty \frac{d}{dt}(S(t) + I(t))\, dt = -\int_0^\infty \gamma I(t) dt.$$

The left-hand side gives

$$\int_0^\infty \frac{d}{dt}(S(t) + I(t))\, dt = S_\infty + I_\infty - S_0 - I_0 = S_\infty - S_0 - I_0,$$

since $I_\infty = 0$. On the other hand, the right-hand side takes the form

$$-\int_0^\infty \gamma I(t) dt = -\gamma \int_0^\infty I(t) dt = -\gamma \hat{I}.$$

We thus have

$$S_\infty - S_0 - I_0 = -\gamma \hat{I}. \tag{6.8}$$

Now consider (6.2a) and divide both sides by S, giving

$$\frac{S'(t)}{S(t)} = -\beta I(t).$$

Integrating both sides from 0 to ∞, we have

$$\ln S_\infty - \ln S_0 = -\beta \hat{I}. \tag{6.9}$$

Express (6.8) and (6.9) in terms of $-\hat{I}$ and equate

$$\frac{\ln S_\infty - \ln S_0}{\beta} = \frac{S_\infty - S_0 - I_0}{\gamma}.$$

Thus, we have

$$\frac{\ln S_\infty - \ln S_0}{\beta} = \frac{S_\infty - S_0 - I_0}{\gamma} \iff \ln S_0 - \ln S_\infty = (S_0 - S_\infty + I_0)\frac{\beta}{\gamma}$$

$$\iff \ln S_0 - \ln S_\infty = (S_0 - S_\infty + I_0)\frac{\beta}{\gamma}\frac{S_0}{S_0}$$

$$\iff \ln S_0 - \ln S_\infty = (S_0 - S_\infty + I_0)\frac{\mathcal{R}_0}{S_0}$$

$$\iff (\ln S_0 - \ln S_\infty)S_0 = (S_0 - S_\infty)\mathcal{R}_0 + I_0\mathcal{R}_0.$$

We can then state the following theorem.

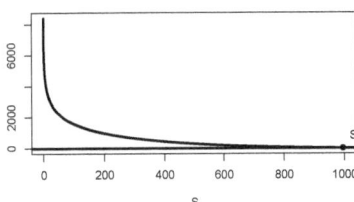 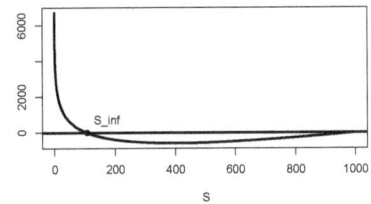

FIGURE 6.6
The function $T(S)$ for $\mathcal{R}_0 = 0.8$ (left) and $\mathcal{R}_0 = 2.4$ (right), for a total population of 1,000.

Theorem 6.2 (Final size relation) *Let $(S(t), I(t))$ be a solution to (6.2) and $\mathcal{R}_0$ be defined by (6.7). The number $S(t)$ of susceptible individuals is a nonincreasing function and its limit S_∞ is the only solution in $(0, S_0)$ of the transcendental equation*

$$(\ln S_0 - \ln S_\infty)S_0 = (S_0 - S_\infty)\mathcal{R}_0 + I_0\mathcal{R}_0. \qquad (6.10)$$

Proof *Rewrite the final size equation (6.10) as*

$$T(S_\infty) := (\ln S_0 - \ln S_\infty)S_0 - (S_0 - S_\infty)\mathcal{R}_0 - I_0\mathcal{R}_0. \qquad (6.11)$$

We thus seek the zeros of the function $T(S_\infty)$, i.e., S_∞ in $(0, S_0]$ such that $T(S_\infty) = 0$. Note to begin that T is a continuous function on $\mathbb{R}_+ \setminus \{0\}$. Furthermore,

$$\lim_{S_\infty \to 0} T(S_\infty) = \lim_{S_\infty \to 0} -S_0 \ln(S_\infty) = \infty.$$

Differentiating T with respect to S_∞, we get

$$T'(S_\infty) = \mathcal{R}_0 - S_0/S_\infty.$$

When $S_\infty \to 0$, $\mathcal{R}_0 - S_0/S_\infty < 0$, so T decreases on the interval $S_\infty \in (0, S_0/\mathcal{R}_0)$ and increases for $S_\infty > S_0/\mathcal{R}_0$. Consequently if $\mathcal{R}_0 \leq 1$, the function T is decreasing on $(0, S_0)$, while it has a minimum if $\mathcal{R}_0 > 1$.
    Case $\mathcal{R}_0 \leq 1$. We have seen that T decreases on $(0, S_0]$. Also, $T(S_0) = -I_0\mathcal{R}_0 < 0$ ($I_0 = 0$ is trivial and not considered). Therefore, by the Intermediate Value Theorem, there exists a unique $S_\infty \in (0, S_0]$ such that $T(S_\infty) = 0$.
    Case $\mathcal{R}_0 > 1$. We have seen that T decreases on $(0, S_0/\mathcal{R}_0]$. For $S_\infty \in [S_0/\mathcal{R}_0]$, $T' > 0$. As before, $T(S_\infty) = -I_0\mathcal{R}_0$. Thus, there exists a unique $S_\infty \in (0, S_0]$ such that $T(S_\infty) = 0$. More precisely, in this case, $S_\infty \in (0, S_0/\mathcal{R}_0)$.

    Figure 6.6 shows the function $T(S)$ for $\mathcal{R}_0 = 0.8$ and $\mathcal{R}_0 = 2.5$, illustrating the two cases in the proof of Theorem 6.2. Figure 6.7 shows the attack rate as a

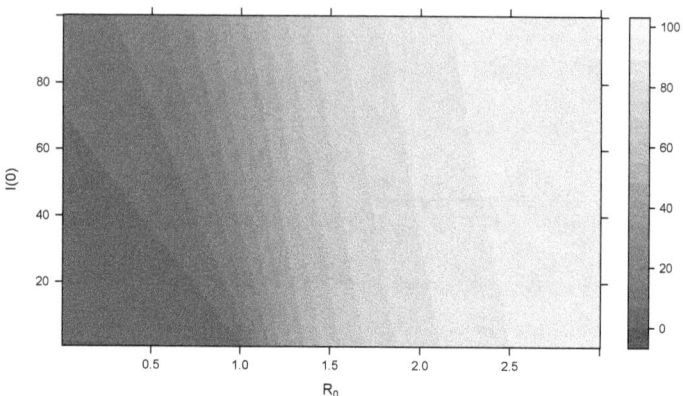

FIGURE 6.7
Attack rate (in %) as a function of $\mathcal{R}_0$ and the initial number of infected individuals $I(0)$.

function of R_0 and I_0. The attack rate is the proportion of the population that has been infected by the end of the epidemic and is obtained by computing $(S_0 - S_\infty)/N_0$.

6.2.5   Herd immunity

To implement vaccination in the Kermack-McKendrick model, assume that vaccination reduces the number of susceptible individuals by moving them to the recovered compartment. Let the total population be N, with S_0 individuals initially susceptible. Suppose we vaccinate a fraction $p \in [0, 1]$ of the susceptible individuals.

Suppose for simplicity that $R(0) = 0$ in the absence of vaccination. The initial condition then takes the form

$$(S(0), I(0), R(0)) = (S_0, I_0, 0).$$

Vaccinating a fraction p of the susceptible population, the initial condition becomes

$$(S(0), I(0), R(0)) = ((1 - p)S_0, I_0, pS_0).$$

Without vaccination, the basic reproduction number is given by (6.7). With vaccination, denoting $\mathcal{R}_v$ the reproduction number, we have

$$\mathcal{R}_v = \frac{\beta}{\gamma}(1 - p)S_0.$$

Since $p \in [0, 1]$, $\mathcal{R}_v \leq \mathcal{R}_0$, i.e., vaccination always reduces the basic reproduction number, with $\mathcal{R}_v < \mathcal{R}_0$ if $p > 0$.

To control the disease, $\mathcal{R}_v$ must take a value less than 1. And we have

$$\mathcal{R}_v < 1 \iff p > 1 - \frac{1}{\mathcal{R}_0}.$$

Thus, by vaccinating a fraction $p > 1 - 1/\mathcal{R}_0$ of the susceptible population, we are in a situation where an epidemic peak is precluded (or, at the very least, the final size is reduced). This is herd immunity.

6.3 The Endemic SIRS Model with Demography

6.3.1 The model(s)

The Kermack-McKendrick model assumes that there is no demographic component to the disease dynamics and focuses on the infectious period of individuals. Obvious modifications of the model include the addition of demographic processes or more focus on other stages of disease transmission. For instance, we could assume that individuals do not die from the disease; after recovering, they become *immune* from infection for some time, then lose immunity and become susceptible again. We can of course combine both! Resulting flow diagrams are shown in Figure 6.8.

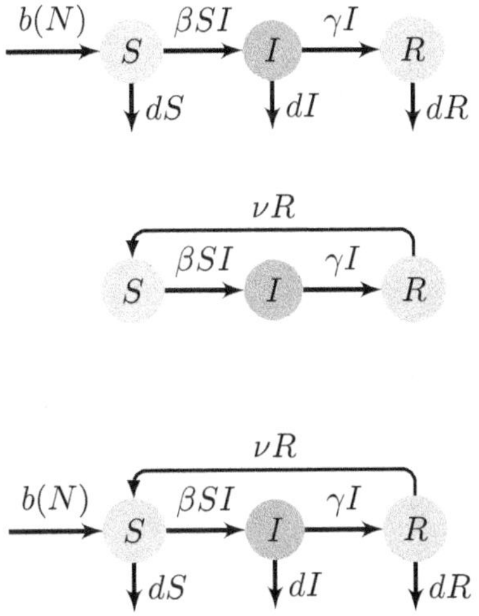

FIGURE 6.8
Three potential variations on the Kermack-McKendrick model. Top: demographic processes are incorporated. Middle: immunity is temporary. Bottom: both demographic processes and temporary immunity are taken into account.

Let us choose the most general model, that with the flow diagram shown in the bottom of Figure 6.8. The system of ordinary differential equations is given by

$$S' = b(N) + \nu R - dS - \beta SI, \tag{6.12a}$$
$$I' = \beta SI - (d + \gamma)I, \tag{6.12b}$$
$$R' = \gamma I - (d + \nu)R. \tag{6.12c}$$

Consider the initial value problem consisting in (6.12) to which we adjoin initial conditions $S(0) = S_0 \geq 0$, $I(0) = I_0 \geq 0$ and $R(0) = R_0 \geq 0$. Typically, we assume $N_0 = S_0 + I_0 + R_0 > 0$ to avoid a trivial case.

Birth and death are *relative* – Remark that the notions of *birth* and *death* are relative to the population under consideration. For instance, consider a model for human immunodeficiency virus (HIV) in an at-risk population of intravenous drug users. Then

- birth is the moment the at-risk behaviour starts;

- death is the moment the at-risk behaviour stops, whether from "real death" or because the individual stops using drugs.

Choosing a form for demography – Before we proceed with the analysis proper, let us discuss the nature of the assumptions on demography. To do this, we consider the behaviour of the total population $N(t) = S(t) + I(t) + R(t)$. Summing the equations in (6.12), we obtain

$$N' = b(N) - dN. \tag{6.13}$$

There are three common ways to define $b(N)$ in (6.13):

1. $b(N) = b$;
2. $b(N) = bN$;
3. $b(N) = bN - cN^2$.

Case 3 leads to logistic dynamics of the total population and is not discussed here. If $b(N) = bN$ (case 2), then birth in (6.13) satisfies $N'/N = b$; we say that birth is constant *per capita*. In this case, (6.13) takes the form

$$N' = bN - dN = (b - d)N,$$

with initial condition $N(0) = N_0$. The solution to this scalar autonomous ordinary differential equation is easy. We obtain

$$N(t) = N_0 e^{(b-d)t}, \quad t \geq 0.$$

Thus there are three possibilities:

- if $b > d$, then $N(t) \to \infty$, i.e., the total population explodes;

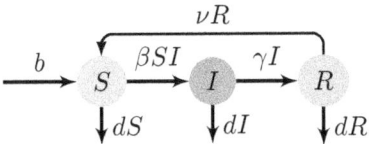

FIGURE 6.9
Flow diagram for the SIRS model under consideration.

- if $b = d$, then $N(t) \equiv N_0$, i.e., the total population remains constant;

- if $b < d$, then $N(t) \to 0$, i.e., the total population collapses.

We want a reasonable case, we could therefore suppose that $b(N) = dN$, which would lead to a constant total population.

However, this is a little reductive, so we choose instead $b(N) = b$ (case 1), which works as well even though it can initially be thought of as not being very realistic since the birth rate, in this case, does not depend on the total population. That is, a population of $N = 1,000$ would for instance have the same number of births per unit time as a population of $N = 1,000,000$.

In this case, (6.13) becomes $N' = b - dN$. This equation can be solved explicitly but is also very easy to consider qualitatively. The equilibrium is $N^\star = b/d$ and it is very easy to check that all solutions, regardless of initial conditions, converge to $N^\star$. Indeed, $N' = b - dN$ is an autonomous scalar equation, so all its solutions are monotone.

Under this assumption on demography, the model has the flow diagram shown in Figure 6.9 and the system of ordinary differential equations describing the evolution of the disease takes the form

$$S' = b + \nu R - dS - \beta SI, \tag{6.14a}$$
$$I' = \beta SI - (d + \gamma)I, \tag{6.14b}$$
$$R' = \gamma I - (d + \nu)R, \tag{6.14c}$$

Consider the initial value problem consisting in (6.14) to which we adjoin initial conditions $S(0) = S_0 \geq 0$, $I(0) = I_0 \geq 0$, and $R(0) = R_0 \geq 0$. Typically, we assume $N_0 = S_0 + I_0 + R_0 > 0$ to avoid a trivial case.

6.3.2 Mathematical analysis

Is the system well-posed? – For an ordinary differential equations epidemiological model, the following questions are typically asked:

- Do solutions to (6.14) exist, and are they unique?

- Is the positive cone (or positive orthant) $\mathbb{R}_+^3$ invariant under the flow of (6.14)?

- Are solutions to (6.14) bounded? Some models have unbounded solutions, but they are rare and need special considerations.

Solutions exist and are unique – The vector field in (6.14) is C^1, implying that solutions exist and are unique. If we had instead considered an incidence of the form $f(S, I, N) = \beta SI/N$ and, say, demography with $b(N) = bN$, then some discussion might have been needed if $b < d$.

Invariance of $\mathbb{R}_+^3$ under the flow – Let us start by assuming that $I(0) = I_0 = 0$. Then (6.14b) takes the form $I' = 0$, meaning that the SR-plane (i.e., the set $\{I = 0\}$) is positively invariant under the flow of (6.14). On that plane, (6.14) reduces to

$$S' = b + \nu R - dS, \tag{6.15a}$$
$$R' = -(d + \nu)R. \tag{6.15b}$$

This implies that a solution with $I_0 > 0$ cannot enter the plane $\{I = 0\}$. Indeed, suppose that $I_0 > 0$ and there exists $t_\star > 0$ such that $I(t_\star) = 0$. Then at $(S(t_\star), I(t_\star) = 0, R(t_\star))$, there are two solutions to (6.14): the one we just generated as well as the one governed by (6.15); this contradicts the uniqueness of solutions to (6.14).

Suppose now that $S = 0$. Equation (6.14a) is then

$$S' = b + \nu R > 0.$$

If $S(0) = S_0 > 0$, then $S(t) > 0$ for all t. If, on the other hand, $S_0 = 0$, then $S(t) > 0$ for $t > 0$ small; from what we just saw, this is then also true for all $t > 0$. We say the vector field points *inward*. This implies that S cannot become zero. We proceed the same way for R.

To summarize, for invariance, denoting $\mathbb{R}^\star = \mathbb{R} \setminus \{0\}$,

- if $(S(0), I(0), R(0)) \in \mathbb{R}_+ \times \mathbb{R}_+^\star \times \mathbb{R}_+$, then for all $t > 0$,

$$(S(t), I(t), R(t)) \in (\mathbb{R}_+^\star)^3;$$

- if $(S(0), I(0), R(0)) \in \mathbb{R}_+ \times \{0\} \times \mathbb{R}_+$, then for all $t \geq 0$,

$$(S(t), I(t), R(t)) \in \mathbb{R}_+^\star \times \{0\} \times \mathbb{R}_+.$$

The model is therefore satisfactory in that it does not allow solutions to become negative.

The total population is asymptotically constant – Since $b(N) = b$, the total population equation (6.13) takes the form

$$N' = b - dN.$$

This equation has a unique equilibrium $N^\star = b/d$ and it is very easy to check that this equilibrium is globally asymptotically stable for solutions with

$N(0) \geq 0$: this is a scalar autonomous equation, so solutions are monotone; they increase to N^* if $N_0 < N^*$ and decrease to N^* if $N_0 > N^*$. So we can work at the limit N^* where $R = N^* - (S+I)$ and thus drop the equation for R.

Boundedness – It follows from what we just saw that the positive cone $\mathbb{R}^3_+$ is (positively) invariant under the flow of (6.14). Since $N(t) \to N^*$, we deduce that solutions of (6.14) are bounded.

Seeking equilibria – We seek $S = S^*$, $I = I^*$, and $R = R^*$ such that

$$0 = b + \nu R^* - dS^* - \beta S^* I^*, \tag{6.16a}$$
$$0 = \beta S^* I^* - (d + \gamma)I^*, \tag{6.16b}$$
$$0 = \gamma I^* - (d + \nu)R^*. \tag{6.16c}$$

From (6.16b), either $I^* = 0$ or $\beta S - (d + \gamma) = 0$, i.e., $S^* = (d + \gamma)/\beta$. When $I^* = 0$, substituting $I^* = 0$ into (6.16c) implies that $R^* = 0$ and, in turn, substituting $I^* = R^* = 0$ into (6.16c) gives $S^* = b/d$. This gives the disease-free equilibrium (DFE)

$$\boldsymbol{E}_0 := (S^*, I^*, R^*) = \left(\frac{b}{d}, 0, 0\right). \tag{6.17}$$

We return to the case where $S^* = (d + \gamma)/\beta$ in a while, for now, we focus on the disease-free equilibrium $\boldsymbol{E}_0$.

Classic method for computing $\mathcal{R}_0$ – $\mathcal{R}_0$ is the surface in parameter space where the disease-free equilibrium loses its local asymptotic stability. To find $\mathcal{R}_0$, we therefore study the disease-free equilibrium. At an arbitrary point (S, I, R), the Jacobian matrix of (6.14) takes the form

$$J_{(S,I,R)} = \begin{pmatrix} -d - \beta I & -\beta S & \nu \\ \beta I & \beta S - (d + \gamma) & 0 \\ 0 & \gamma & -(d + \nu) \end{pmatrix}. \tag{6.18}$$

Local asymptotic stability of the disease-free equilibrium $\boldsymbol{E}_0$ depends on the sign of the real parts of the eigenvalues of (6.18) at that equilibrium point, so we evaluate

$$J_{\boldsymbol{E}_0} = \begin{pmatrix} -d & -\beta S^* & \nu \\ 0 & \beta S^* - (d + \gamma) & 0 \\ 0 & \gamma & -(d + \nu) \end{pmatrix}. \tag{6.19}$$

This is a block upper triangular matrix, with the second diagonal block a lower triangular matrix, implying its eigenvalues are $-d < 0$, $-(d + \nu) < 0$, and $\beta S^* - (d + \gamma)$. Thus, the local asymptotic stability of the disease-free equilibrium is determined by the sign of $\beta S^* - (d + \gamma)$.

Recall that at the disease-free equilibrium (6.17), $S^\star = b/d$, so

$$\text{sign}(\beta S^\star - (d + \gamma)) = \text{sign}\left(\beta \frac{b}{d} - (d + \gamma)\right).$$

So the disease-free equilibrium is locally asymptotically stable if $\beta b/d < d + \gamma$, or, in other words, if

$$\frac{\beta}{d + \gamma} \frac{b}{d} < 1.$$

Denote

$$\mathcal{R}_0 = \frac{\beta}{d + \gamma} \frac{b}{d}.$$

We often emphasize that $b/d = N^\star$, the total population, and thus write $\mathcal{R}_0 = \beta N^\star / (d + \gamma)$.

Now consider the second EP where $S^\star = (d + \gamma)/\beta = N^\star / \mathcal{R}_0$. Write (6.16c) as $R^\star = \gamma I^\star / (d + \nu)$. Since $S^\star + I^\star + R^\star = N^\star$, this means that

$$N^\star - S^\star - I^\star = \gamma I^\star / (d + \nu),$$

so substituting $S^\star = N^\star / \mathcal{R}_0$,

$$\left(1 + \frac{\gamma}{d + \nu}\right) I^\star = \left(1 - \frac{1}{\mathcal{R}_0}\right) N^\star.$$

So finally,

$$I^\star = \left(1 - \frac{1}{\mathcal{R}_0}\right) \frac{d + \nu}{d + \nu + \gamma} N^\star.$$

The endemic equilibrium (EEP) of (6.14) is

$$\boldsymbol{E}_\star := (S^\star, I^\star, R^\star) = \left(\frac{1}{\mathcal{R}_0} N^\star, \left(1 - \frac{1}{\mathcal{R}_0}\right) \frac{d + \nu}{d + \nu + \gamma} N^\star, N^\star - (S^\star + I^\star)\right).$$
$$(6.20)$$

Remark that $\boldsymbol{E}_\star$ is not biologically relevant when $\mathcal{R}_0 \leq 1$. By that, we mean that the equilibrium is not in the positive cone $\mathbb{R}_+^3$, so it is not a physically meaningful state of the system. So far, we have proved the following result.

Theorem 6.3 *Let the basic reproduction number be*

$$\mathcal{R}_0 = \frac{\beta}{d + \gamma} N^\star \tag{6.21}$$

and consider the equilibria of (6.14): the disease-free equilibrium (6.17) and the endemic equilibrium (6.20).

- *If $\mathcal{R}_0 < 1$, then $\boldsymbol{E}_0$ is locally asymptotically stable and $\boldsymbol{E}_\star$ is not biologically relevant.*

- *If $\mathcal{R}_0 > 1$, then $\boldsymbol{E}_0$ is unstable and $\boldsymbol{E}_\star$ is biologically relevant.*

As you can probably guess, if $\mathcal{R}_0 > 1$, then $\boldsymbol{E}_\star$ is not only biologically relevant but also locally asymptotically stable. Recall the Jacobian matrix of (6.14) at an arbitrary point (S, I, R) is given by (6.18). Let us rewrite it as

$$J_{(S,I,R)} = \begin{pmatrix} -\beta I & -\beta S & \nu \\ \beta I & \beta S - \gamma & 0 \\ 0 & \gamma & -\nu \end{pmatrix} - d\, \mathbb{I}.$$

From this, we get that $-d$ is an eigenvalue of J:

- there is a theorem that tells us that if $\lambda \in \sigma(M)$, then $\lambda + k \in \sigma(M + k\mathbb{I})$, where $\sigma(M)$ is the spectrum of M, the set of eigenvalues of M;

- the first matrix has all column sums zero so has a zero eigenvalue.

The same way, the remaining two eigenvalues of $J_{\boldsymbol{E}_\star}$ can be found by considering the matrix

$$\begin{pmatrix} -\beta I^\star & -\beta S^\star & \nu \\ \beta I^\star & \beta S^\star - \gamma & 0 \\ 0 & \gamma & -\nu \end{pmatrix}$$

and shifting its spectrum by $-d$. We find

$$-\frac{\sqrt{\nu^2 + 2(\beta(S^\star - I^\star) - \gamma)\nu + \gamma^2 - 2\beta\gamma(S^\star + I^\star) + (S^\star - I^\star)^2\beta^2} + \nu + \gamma + (I^\star - S^\star)\beta}{2}$$

and

$$\frac{\sqrt{\nu^2 + 2(\beta(S^\star - I^\star) - \gamma)\nu + \gamma^2 - 2\beta\gamma(S^\star + I^\star) + (S^\star - I^\star)^2\beta^2} - \nu - \gamma - (I^\star - S^\star)\beta}{2}.$$

We could continue and after some blood, sweat and tears, get that $J_{\boldsymbol{E}_\star}$ has its eigenvalues with negative real parts when $\boldsymbol{E}_\star$ is biologically relevant, i.e., when $\mathcal{R}_0 > 1$. With even more blood, sweat, and tears, we can actually show that the result is *global*. We express that in the next result.

Theorem 6.4 *Let the basic reproduction number be defined by* (6.21) *and consider the disease-free equilibrium* (6.17) *and the endemic equilibrium point* (6.20),

- *If* $\mathcal{R}_0 < 1$, *then* $\boldsymbol{E}_0$ *is globally asymptotically stable and* $\boldsymbol{E}_\star$ *is not biologically relevant.*

- *If* $\mathcal{R}_0 > 1$, *then* $\boldsymbol{E}_0$ *is unstable and* $\boldsymbol{E}_\star$ *is globally asymptotically stable.*

In other words:

- when $\mathcal{R}_0 < 1$, then all solutions go to the DFE, and the disease goes extinct.

- when $\mathcal{R}_0 > 1$, then all solutions go to the EEP, and the disease becomes endemic.

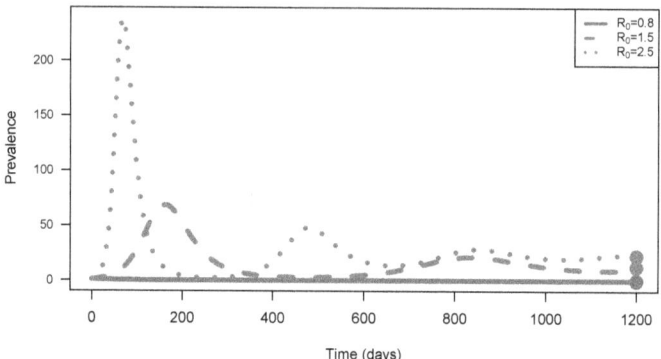

FIGURE 6.10
A few solutions of the SIRS model.

6.3.3 Some numerics

We now illustrate the results of Theorems 6.3 and 6.4 with some numerical simulations. Figure 6.10 shows the prevalence of the disease for three different values of $\mathcal{R}_0$. We could continue, but with a model this simple, there is little more to do: the 3 parameters of the system are combined within $\mathcal{R}_0$, and the latter summarizes the dynamics well.

Let us now show an important summary of the situation using a so-called bifurcation diagram. We saw that when $\mathcal{R}_0 < 1$, $I(t) \to 0$, whereas when $\mathcal{R}_0 > 1$, $I \to I^\star$. Let us represent this in Figure 6.11. The bifurcation diagram shows the prevalence at the endemic equilibrium as a function of $\mathcal{R}_0$. When $\mathcal{R}_0 < 1$, the endemic equilibrium is not biologically relevant, so we show the prevalence at the disease-free equilibrium, i.e., $I = 0$. When $\mathcal{R}_0 > 1$, the endemic equilibrium is biologically relevant, so we show the prevalence at the endemic equilibrium. The stability of the equilibria is also shown: a continuous line means the equilibrium is locally asymptotically stable, whereas a dashed line means the equilibrium is unstable.

6.4 Last Remarks

To simplify or not to simplify? In the Kermack-McKendrick epidemic model (6.1) and the SIRS endemic model (6.14), since the total population is constant or asymptotically constant, it is possible to omit one of the state variables since $N^\star = S + I + R$. We often use $R = N^\star - S - I$. This can greatly simplify some computations. Whether to do it or not is a matter of preference. But one should bear in mind that sometimes, things "get lost" when simplifying.

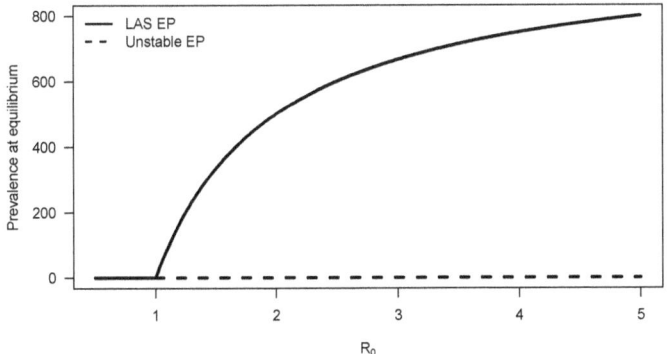

FIGURE 6.11
Bifurcation diagram of the SIRS model.

To normalize or not to normalize? In the Kermack-McKendrick epidemic model (6.1) and the SIRS endemic model (6.14), since the total population is constant or asymptotically constant, it is possible to work with $N = 1$. This can greatly simplify some computations. It is important to always have the "sizes" of objects in mind. If you do normalize, always do a "return to biology", i.e., interpret your results in a biological light, which often implies to return to original values.

Where we are – we have seen an *epidemic* SIR model (the Kermack-McKendrick SIR) in which the presence or absence of an epidemic wave is characterized by the value of $\mathcal{R}_0$. The Kermack-McKendrick SIR has explicit solutions (in some sense). *This is an exception!* In practice, it is extremely rare to encounter models with explicit solutions; analysis using qualitative methods is preferable and often the only way to proceed.

We also saw an *endemic* SIRS model in which the threshold $\mathcal{R}_0 = 1$ is such that, when $\mathcal{R}_0 < 1$, the disease goes extinct, whereas when $\mathcal{R}_0 > 1$, the disease becomes established in the population.

Extensions – We barely touched the surface of epidemiological models. The literature is extremely abundant and varied. Some examples of extensions are given as exercises, but they, too, are just the tip of the iceberg.

6.5   Exercises

6.5.1   An SIRS model with vaccination

Take SIRS model (6.14) and assume the following:

- vaccination takes newborn individuals and moves them directly into the removed compartment, without them becoming infected/infectious;

- A fraction p is vaccinated at birth.

Thus, the model has the following flow diagram.

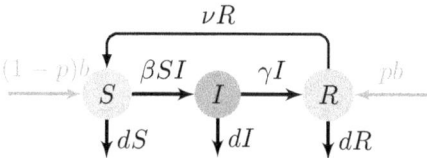

Write and analyse the resulting model. You should observe that the disease-free equilibrium now comprises a positive value for $R^\star$. However, the remainder of the analysis should follow the same lines as the original SIRS model (6.14). Compute the reproduction number $\mathcal{R}_v$ with vaccination and interpret its value in terms of herd immunity: what must the value of p be so that $\mathcal{R}_v < 1$, in terms of the basic reproduction number $\mathcal{R}_0$ of the original SIRS model (6.14)?

6.5.2 An SIR model with immigration of infectious individuals

Consider an SIR model with a constant influx of individuals into the population as well as a constant outflow of individuals from the population, at the same rate m as that of the inflow. Thus the flow diagram is as follows.

Write and try to analyse the model. You should observe that there is no disease-free equilibrium, and as a consequence, it is impossible to compute a reproduction number.

6.5.3 An SLIRS model

The models considered this far assume that individuals move directly from the susceptible to the infectious compartment. However, many disease have an incubation period during which individuals are infected but not yet infectious but are already infected. To model this, we typically use an SLIR, where the L compartment represents individuals during the latent phase of the disease. Historically, such models have been called the SEIR model, with the E standing for "exposed". However, the term "exposed" is misleading as it implies that individuals are not yet infected, which is not the case.

The SLIRS model has the following flow diagram.

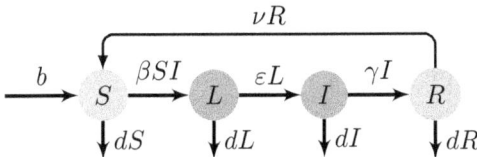

Write the model and analyze it. You will probably only be able to compute the reproduction number $\mathcal{R}_0$ and the local asymptotic stability of the disease-free equilibrium, but if you are feeling adventurous, you can try to compute the endemic equilibrium and study its local stability.

7

Modelling based on Conservation Principles

In the rest of this book, we shall consider concepts and approaches for models based on *conservation principles*. Some of the material will be relatively easy, at least for students with some background in fluid and continuum mechanics. Nevertheless, it is important to see that the principles are generally valid and applicable for modelling in many other contexts than traditional mechanics.

It is difficult to discuss conservation principles without getting involved with partial differential equations, but we shall mostly be interested in the qualitative theory and the general behaviour of the solutions, and not very specific analytical or numerical solution methods. Below, the term *fluid* is used both for liquids and gases. A *first aid* course about first-order quasi-linear partial differential equations is found in Chapter 12. It is designed for readers without any knowledge of partial differential equations beyond calculus.

In many models we apply continuous distributions or densities in space where the phenomenon we consider is discrete. Typical examples could be the density of bacteria, people, cars, and the like. In such cases, the models implicitly assume a kind of *continuum hypothesis* similar to the one in continuum mechanics. However, such an approach has obvious limitations, and it is important to be aware of what the models are really good for. It leads for instance to absurdities to insist on mathematical definitions based on limits to ∞ or 0. In the physical world "0" is a few orders of magnitudes below and "∞" some orders of magnitudes above where we are located. When we say that "$\Delta x \to 0$", we actually mean that Δx is small compared to the scale where we are located, and not that Δx really goes to 0 in the mathematical sense. This is similar to talking about stationary conditions in time from $-\infty$ and ∞. In elementary particle physics the eternity could well be 10^{-10} s! Scaling considerations that we have covered earlier, help us to assess the reliability of our assumptions.

We should distinguish between establishing a model and solving the equations after they have been formulated in a mathematical model. The latter is the theme of the courses in analytical and numerical solution of ordinary and partial differential equations. Although enthusiasts advertise numerical software that can solve any differential equation, we are still far from leaving to the computer to *understand* what is really happening. As numerical

DOI: 10.1201/9781003725206-7

tools are becoming more advanced, it is, on the contrary, an increasing demand for mathematical expertise and analysis of the equations. Today, serious customers require that calculations based on numerical models should be documented to be reliable. Only thorough mathematical and numerical analysis, and not least physical and engineering insight, can help with this. Many of the analytical solutions for idealized problems that are known from the theory of partial differential equations are useful in this respect. Scaling arguments show that the so-called fundamental solution of the heat conduction equation has far greater applicability than is usually mentioned in the mathematics courses. In a way, fundamental solutions and other solutions from idealized mathematical situations are the cornerstones that give us insight and set limits.

7.1 Basic Concepts

7.1.1 Density

Although we perceive water and air as quite homogeneous and uniform physical materials, we all know that this is only when considered from our own length-scale. If we made an imaginary sphere with radius r and centre at $\mathbf{x}$ in air and could calculate the mass within the sphere, $m(r)$, the mathematical definition of the density of the air at the point $\mathbf{x}$ would be

$$\rho(\mathbf{x}) = \lim_{r \to 0} \frac{m(r)}{4\pi r^3/3} \tag{7.1}$$

If we were really able to perform this experiment, and plot the ratio in a graph as a function of r, we would, however, see something like in Figure 7.1. When r becomes less than $r_{\min}$ (about 10^{-7}m for the air around us), the ratio begins to fluctuate, and it certainly makes no sense in talking about a limit when r goes to 0, as we do in mathematics. Conversely, if r is too large, the ratio will no longer be constant because the air inside the sphere is no longer uniform. As you understand, we must add to the definition of density the assumption that we stop the limit process in the right place, and that our definition of the density of air only has meaning for phenomena with a length scale between $r_{\min}$ and $r_{\max}$. As applied mathematicians, we have to bear with density not being particularly "well-defined". This does not create major problems for air and water in most of our daily situations, but for high vacuum technology, $r_{\min}$ may well be of the same scale as the apparatus.

Let us consider some quantity that we describe by a density $\varphi(\mathbf{x}, t)$. The amount within a given closed region R of space may be expressed by the volume integral,

$$M(t) = \int_R \varphi(\mathbf{x}, t) \, dV.$$

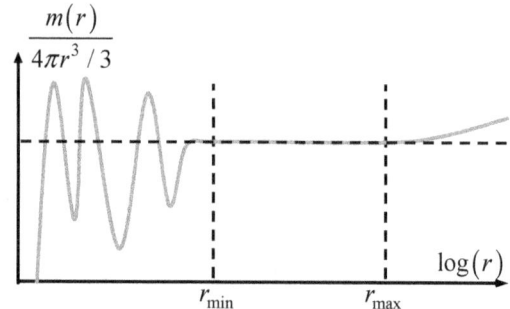

FIGURE 7.1

Pragmatic definition of *density* at a point in space: The mass over volume ratio is reasonably stable for $r_{min} < r < r_{max}$, but the limit when $r \to 0$ does not exist and makes no sense.

Although one immediately thinks of density as amount per volume unit, there is nothing wrong in defining the density as the amount per area unit, like 80 g/m^2 for ordinary writing paper. Similarly, for a thin iron wire, the useful density becomes mass per unit length.

In the introduction to [30], the authors have two examples which illustrate how it is possible to think of continuous densities in two quite extreme cases. One example discusses how the spiral structure of galaxies can be modelled as density waves in a gravitational plasma in which the galaxies are modelled as a continuum characterized by the mass density of the stars (C. C. Lin was in fact one of the main contributors to this theory). The second example discusses instabilities in the density of amoebas during food shortages, where the density is the number of amoebas per area unit, regarded as a continuous function of the position on the surfaces where they live (this model and the analysis of it were developed by the second author, L. A. Segel).

In mathematical modelling we therefore talk about densities in many other situations than those known from mechanics. The density of various foreign substances, such as contaminants in water, is also a relevant example. Within the air pollution modelling, the most advanced mathematical models consider hundreds of different components, each of which is characterized by its density. In addition, the components interact, decay chemically, are transported with the wind and become mixed in the air masses, or simply fall down. Oil reservoir engineering applies complicated mathematical models for tracing various oil and gas components in porous rocks.

The heat or energy content in materials may be expressed as energy per unit volume. This will in the simplest case with constant specific heat be proportional to the temperature. Entropy density appears in models that deal with heat conduction and heat transfer. Some densities lead us into mathematical problems (so-called *singular densities*) which we shall return to below in the section about sources and sinks.

In continuum mechanics, quantities that passively follow the flow are called *material variables*. The most common material variables in mechanics are mass, momentum, vorticity and energy, which, in a continuous medium, are described by

ρ	mass density
$\rho \mathbf{v}$	momentum density
$\nabla \times \rho \mathbf{v}$	Vorticity density
$e\rho$	energy density

where $\mathbf{v}$ is the velocity of the medium. Impurities or other additives that passively follow the flow, are also material variable. The concentration of plankton in the water is therefore a material variable as long as it does not move on its own. On the other hand, a school of herring is definitely not a material variable!

Within biology it is common, as in the example by L. A. Segel, to operate with continuous density functions of animals, bacteria and plants. This allows us to create models describing the motion of animal herds, bacterial cultures, the spread of epidemics, and the like. We shall later look at a situation where we model the density of cars along a road as a continuum.

7.1.2 Flux

Flux is about transport or flow of something. The term has actually various meanings in science, but here it is only connected to the motion. If we stand by a road watching the cars passing, the average number of cars passing per minute will be what we define as the *flux of cars*. Flux includes the direction of the flow, so the flux of cars should be separated into flux to the right, and flux to the left.

We will meet flux in many different situations. To fix ideas, let us consider the flow of some material through space. Standing at a fixed point $\mathbf{x}$ we observe that material passes, but in order to quantify how much, we put (an imaginary) open small window frame $d\sigma$ into the stream at $\mathbf{x}$ and observe how much is passing through the frame per second. It is convenient to present the measurements as an amount per second and area unit, since a window twice as large and with the same orientation would allow twice as much material to pass through. The orientation of the frame is uniquely defined by a normal vector $\mathbf{n}$ attached to the frame. If we change the orientation and hence the normal vector to the opposite direction, the flux changes sign. The maximum amount will flow through the frame if we align $\mathbf{n}$ with the direction of the flow, and this direction is what we define to be the *direction of the flux*. Flux can therefore most easily be described as a *vector field*, $\mathbf{j}(\mathbf{x}, t)$, where the direction of $\mathbf{j}$ indicates the transport direction, and the size, $|\mathbf{j}|$, the amount per time and area unit. Nothing passes through the frame if $\mathbf{n}$ is orthogonal to $\mathbf{j}$. In general, the amount dM that passes through $d\sigma$ (with orientation $\mathbf{n}$) during a time period dt is thus

$$dM = \mathbf{j} \cdot \mathbf{n} d\sigma dt. \tag{7.2}$$

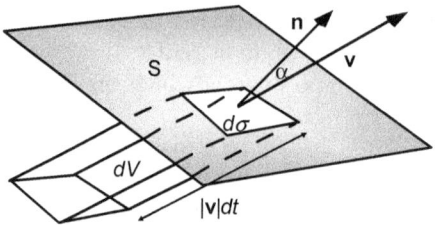

FIGURE 7.2
Derivation of the expression for the flux of a material variable.

The total amount flowing out through a surface Σ in space with normal vector **n** per unit of time is now given by the surface integral,

$$\frac{dM}{dt} = \int_{\Sigma} \mathbf{j} \cdot \mathbf{n} d\sigma.$$

By means of the *Divergence Theorem* from vector analysis, it is possible to rewrite the surface integrals over the closed surface ∂R of a volume R as

$$\int_{\partial R} \mathbf{j} \cdot \mathbf{n} d\sigma = \int_{R} \nabla \cdot \mathbf{j} dV. \tag{7.3}$$

We shall later see that some care must be exercised when applying the Divergence Theorem.

For a material variable with density φ, passively following a continuous flow with a velocity vector field **v**, the flux has a particularly simple form, namely, $\mathbf{j} = \varphi \mathbf{v}$. This follows from Figure 7.2, where an amount φdV passes through $d\sigma$ during the time dt. Simple vector calculus gives $dV = d\sigma \cdot |\mathbf{v}| dt \cdot \cos \alpha = \mathbf{v} \cdot \mathbf{n} d\sigma dt$. Therefore, the amount passing through $d\sigma$ per time unit will be

$$Q = \frac{\varphi dV}{dt} = \varphi \mathbf{v} \cdot \mathbf{n} d\sigma = \mathbf{j} \cdot \mathbf{n} d\sigma. \tag{7.4}$$

The flux for this particular case may be written as $\mathbf{j} = \varphi \mathbf{v}$, and the total flow of the material through a surface Σ per time unit is

$$Q = \int_{\Sigma} (\varphi \mathbf{v}) \cdot \mathbf{n} d\sigma. \tag{7.5}$$

There exist a lot of different expressions for flux. In a practical modelling situation it may sometimes be difficult to come up with a good model. The expression for the turbulent dispersion and transport of material considered later in this note is not yet fully resolved, despite more than 100 years of active research. In electricity, we have in the general form of Ohm's law that the flux of electric current, **j**, is given by $\mathbf{j} = \sigma \mathbf{E}$ where σ is the material conductivity and **E** electric field strength. In electromagnetism, where electromagnetic radiation carries energy, the energy flux is given by the Poynting vector, $\mathbf{P} = \mathbf{E} \times \mathbf{H}$.

We mentioned in the previous section that plankton could be a material variable. Now it is known that plankton to some extent is attracted by light, and therefore establish a flux directed towards the light.

Of the more curious models of flux, we have the assumption that schools of fish tend to move along the gradient of its *well-being* function, g, such that

$$\mathbf{j} \propto \nabla g.$$

We shall later return to other models of flux, for example, diffusion-generated flux.

7.1.3 Sources and sinks

A *source* produces a certain amount of substance per time unit, and we may define a *sink* as a source with negative output, hence we only discuss sources. Mathematically, a study of a source at the point $\mathbf{x}_0$ could be carried out by considering a ball R of radius r around $\mathbf{x}_0$, then computing

$$\int_{\partial R} \mathbf{j} \cdot \mathbf{n} d\sigma, \tag{7.6}$$

and see what happens when $r \to 0$. Again, we must have the same reservations in mind as we had for definition of density.

One usually distinguishes between *distributed* and *singular* sources. For a distributed source, the limit value of

$$\lim_{r \to 0} \frac{\int_{\partial R} \mathbf{j} \cdot \mathbf{n} d\sigma}{4\pi r^3/3} \tag{7.7}$$

exists, and when $\mathbf{x}_0$ varies, it defines a function, a so-called *production density* $q(\mathbf{x}, t)$ that expresses the production per time and volume unit. The production density is related to divergence of flux field if the flux and its divergence are nice and continuous functions. Applying the Divergence and Mean Value Theorems,

$$\int_{\partial R} \mathbf{j} \cdot \mathbf{n} d\sigma = \int_R \nabla \cdot \mathbf{j} dV = (\nabla \cdot \mathbf{j})(\mathbf{x}') \times 4\pi r^3/3, \ |\mathbf{x}' - \mathbf{x}_0| < r, \tag{7.8}$$

and thus,

$$\lim_{r \to 0} \frac{\int_{\partial R} \mathbf{j} \cdot \mathbf{n} d\sigma}{4\pi r^3/3} = \lim_{r \to 0} (\nabla \cdot \mathbf{j})(\mathbf{x}') = (\nabla \cdot \mathbf{j})(\mathbf{x}_0). \tag{7.9}$$

In conclusion,

$$q(\mathbf{x}, t) = \nabla \cdot \mathbf{j}(\mathbf{x})$$

Obviously, for a certain production density, the total production in the volume R during the time from t_1 to t_2, is

$$Q = \int_{t_t}^{t_2} \int_R q(\mathbf{x}, t) \, dV dt. \tag{7.10}$$

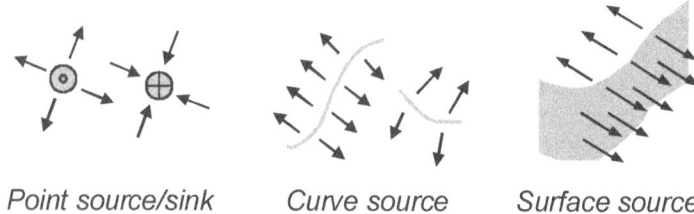

Point source/sink   Curve source   Surface source

FIGURE 7.3
Illustration of various singular sources.

The simplest singular sources are the *point sources*. For point sources we have, as the name indicates, only production from a single point $\mathbf{x}_0$. This is an idealized situation where, in practice, $q(\mathbf{x}, t)$ only differs from zero near $\mathbf{x}_0$. For a ball R around $\mathbf{x}_0$ as considered above,

$$\int_{\partial R} \mathbf{j} \cdot \mathbf{n} d\sigma \qquad (7.11)$$

will stay constant as r tends to 0, or more generally be a function of time. If we were to describe a point source at $\mathbf{x}_0$ by a production density, this would be 0 for $\mathbf{x} \neq \mathbf{x}_0$, while the integral over an arbitrarly small region containing $\mathbf{x}_0$ would be different from 0. Regular functions with this property do not exist, but one has introduced *generalized functions* (also called *distributions*) to be used in such situations. A point source at $\mathbf{x}_0$ can thus be described using the so-called δ-function,

$$q(\mathbf{x}, t) = Q(t) \delta_{\mathbf{x}_0}(\mathbf{x}). \qquad (7.12)$$

Here, $Q(t)$ denotes the production rate (amount per time unit), and the δ-function states that production occurs at point $\mathbf{x}_0$. In general, the δ-function at $\mathbf{x}_0$ is defined by the property that

$$\int f(x) \delta_{\mathbf{x}_0}(\mathbf{x}) dV = f(\mathbf{x}_0)$$

for all continuous functions.

Point sources do not completely cover the category of singular sources. Often, one needs to model sources located along a surface or a curve in the area to be considered. For example, one can think of a varying heat production along a curve (think of an electrical resistance wire!) or a surface, as indicated in Figure 7.3. This also leads to generalized functions, but one can always model such sources as limits of point sources where these spread out along a curve or on a surface. We leave it to the reader to find out how the above integral, $\int_{\partial R} \mathbf{j} \cdot \mathbf{n} d\sigma$, will behave when $r \to 0$ for curve and surface sources.

In practice, it is often appropriate to include also the singular sources in the production density, which then, in the mathematical sense, becomes a generalized function.

From physics (or distribution theory in mathematics), *dipole* and *quadrupole* sources are also known. We will not meet these in any of the situations we shall consider here.

7.2   The Universal Conservation Law

Let us consider a geometrically closed, imaginary region R with boundary ∂R in space. In modelling, such a region is often called a *control volume*. No part of the boundary needs to be physical, so that material may flow freely through.

In space we have a material with density $\rho(\mathbf{x},t)$ moving with a flux, $\mathbf{j}(\mathbf{x},t,\rho)$, which depends on $\mathbf{x}$, t, and ρ. For the sources and sinks inside R, we prescribe a generalized production density $q(\mathbf{x},t)$, which may in general contain singular sources. From what we have been through, we see that the rate of change in the total amount in R may be written

$$\frac{\mathrm{d}}{\mathrm{d}t} \int_R \rho(\mathbf{x},\mathbf{t})\,\mathrm{d}V. \tag{7.13}$$

Now, this must be equal to *minus* what is disappearing over the boundary of R per time unit (because of the definition of an *outer* unit normal, $\mathbf{n}$),

$$-\int_{\partial R} \mathbf{j}(\mathbf{x},t,\rho)\cdot\mathbf{n}\mathrm{d}\sigma \tag{7.14}$$

plus what is produced (or disappears) in R per time unit,

$$\int_R q(\mathbf{x},t)\,\mathrm{d}V. \tag{7.15}$$

Altogether, we obtain the equation

$$\frac{\mathrm{d}}{\mathrm{d}t} \int_R \rho(\mathbf{x},t)\,\mathrm{d}V + \int_{\partial R} \mathbf{j}(\mathbf{x},t,\rho)\cdot\mathbf{n}\mathrm{d}\sigma = \int_R q(\mathbf{x},t)\,\mathrm{d}V. \tag{7.16}$$

This is called a *conservation law* in *integral form*. The integrals and the derivative with respect to t in the first term exist under fairly general conditions. Otherwise, the conservation law is the mathematical formulation that "nothing can disappear or arise from nothing", – a law of nature with an overwhelming empirical basis!

If ρ and $\mathbf{v}$ are sufficiently smooth functions of t and $\mathbf{x}$, we can move the derivative with respect to t under the integral sign, and otherwise use the

Divergence Theorem,

$$\frac{\mathrm{d}}{\mathrm{d}t} \int_R \rho \mathrm{d}V = \int_R \frac{\partial \rho}{\partial t} \mathrm{d}V,$$

$$\int_{\partial R} \mathbf{j} \cdot \mathbf{n} \mathrm{d}\sigma = \int_R \nabla \cdot \mathbf{j} \mathrm{d}V.$$

This means that

$$\int_R \left(\frac{\partial \rho}{\partial t} + \nabla \cdot \mathbf{j} - \mathbf{q} \right) \mathrm{d}V = 0. \tag{7.17}$$

Such a relation would actually hold for all nice R if ρ and $\mathbf{v}$ are smooth and nice functions in the domain we are considering (mathematically, it is sufficient that it holds for all spheres in the domain). If then $\rho_t + \nabla \cdot \mathbf{j} - q$ is continuous, a result from analysis says that

$$\rho_t + \nabla \cdot \mathbf{j} - q \equiv 0 \tag{7.18}$$

in R (sketch of proof: suppose that for a fixed t, $f(\mathbf{x}) = \rho_t + \nabla \cdot \mathbf{j} - q$ is different from 0 at the point $\mathbf{x}_0$ in the interior of R. Then f is different from 0 for all $\mathbf{x}$ a neighbourhood $N \subset R$ around $\mathbf{x}_0$ since f is continuous. Consequently, the integral of f over this neighbourhood is also different from 0, contradictory to the assumption). Equation (7.18) thus applies when ρ, $\mathbf{j}$, $\mathbf{v}$ and q are smooth, and this is the conservation law stated in *differential form*.

Since we will later see examples where one cannot move the derivation inside the integration sign, the integral formulation is more general and fundamental than the differential formulation.

7.3   Conservation Laws in one Space Dimension

We shall discuss some properties of simple conservation laws, and limit ourselves to a simple one-dimensional situation. The *density*, $\rho(x, t)$, has now dimensionless amount per unit length. The *flux* $j(x, t, \rho)$ expresses the amount of material passing the point x (in the positive x-direction) per time unit, and will in general also be a function of t and ρ. Note that the flux is a *vector* directed along the x-axis. In this discussion, we shall for simplicity ignore sources and sinks.

For a finite segment $[A, B]$ of the x-axis, we may, since nothing disappears or is produced in $[A, B]$ write

$$\frac{\mathrm{d}}{\mathrm{d}t} \int_A^B \rho(x, t) \, \mathrm{d}x + j(B, t, \rho(B, t)) - j(A, t, \rho(A, t)) = 0. \tag{7.19}$$

Note that the boundary unit vector $\mathbf{n}$ is $-\mathbf{i}$ in A and $\mathbf{i}$ in B. If $\rho(x, t)$ is a sufficiently smooth function, we have for all possible subintervals $[a, b]$ of

$[A, B]$ that

$$\frac{\mathrm{d}}{\mathrm{d}t} \int_a^b \rho(x, t)\, \mathrm{d}x = \int_a^b \frac{\partial \rho(x, t)}{\partial t} \mathrm{d}x = \frac{\partial \rho(\xi, t)}{\partial t}(b - a), \ \xi \in (a, b). \qquad (7.20)$$

Thus, we may write

$$\frac{\partial \rho(\xi, t)}{\partial t} + \frac{j(b, t, \rho(b, t)) - j(a, t, \rho(a, t))}{b - a} = 0. \qquad (7.21)$$

Using the chain rule and definition of derivative, and in the limit $a \to b$, we obtain the differential equation

$$\frac{\partial \rho}{\partial t} + \frac{\partial j}{\partial x} + \frac{\partial j}{\partial \rho}\frac{\partial \rho}{\partial x} = 0. \qquad (7.22)$$

This is the conservation law in differential form, which mathematically is a (generally nonlinear) hyperbolic partial differential equation.

The following simple (but important!) example illustrates why one can not in general write

$$\frac{\mathrm{d}}{\mathrm{d}t} \int_a^b \rho(x, t)\, \mathrm{d}x = \int_a^b \frac{\partial \rho(x, t)}{\partial t} \mathrm{d}x. \qquad (7.23)$$

Assume solid medium has a discontinuity in the density from ρ_1 to ρ_2. The medium moves with uniform speed U along the x-axis. Since the flux in this case is just ρU, the conservation law with the discontinuity *between* a and b, gives

$$\frac{\mathrm{d}}{\mathrm{d}t} \int_a^b \rho \mathrm{d}x + \rho_2 U - \rho_1 U = 0, \qquad (7.24)$$

or,

$$\frac{\mathrm{d}}{\mathrm{d}t} \int_a^b \rho \mathrm{d}x = (\rho_1 - \rho_2)\, U.$$

However, since $\partial \rho / \partial t$ is equal to 0 except at the discontinuity, say at $x = x_g$, we could also write

$$\frac{\mathrm{d}}{\mathrm{d}t} \int_a^b \rho \mathrm{d}x = \int_a^{x_g} \frac{\partial \rho_1}{\partial t} \mathrm{d}x + \int_{x_g}^b \frac{\partial \rho_2}{\partial t} \mathrm{d}x = 0 + 0 = 0. \qquad (7.25)$$

Consequently,

$$\frac{\mathrm{d}}{\mathrm{d}t} \int_R \rho \mathrm{d}x \neq \int_R \frac{\partial \rho}{\partial t} \mathrm{d}x. \qquad (7.26)$$

This is a situation which may well occur in practice.

Without getting too far into the theory of partial differential equations, we shall limit ourselves to a situation where $\mathbf{j}$ is a known, differentiable function of ρ. The differential equation is then reduced to

$$\frac{\partial \rho}{\partial t} + c(\rho)\frac{\partial \rho}{\partial x} = 0, \quad c(\rho) = \frac{\mathrm{d}j}{\mathrm{d}\rho}. \qquad (7.27)$$

The quantity $c(\rho)$ has dimension velocity, and is called the *kinematic velocity*. In a way which becomes clearer later, one could say that the kinematic speed represents the speed of information in the problem.

To find the solution of this differential equation, we make the following interesting observation: Suppose that the solution $\rho(x, t)$ is already known. Then we also know $c(\rho(x, t))$. Let us define a vector field in the (x, t)-plane by

$$\mathbf{v}(x, t) = \{c(\rho(x, t)), 1\}. \tag{7.28}$$

Field curves (with the curve length parameter s) are defined by the equations

$$\frac{\mathrm{d}x}{\mathrm{d}s} = c(\rho(x, t)),$$

$$\frac{\mathrm{d}t}{\mathrm{d}s} = 1.$$

Assume that $x = p(s)$ and $t = q(s)$, for $-\infty < s < \infty$ represent the curves. We calculate the variation of ρ along a curve by means of

$$\frac{\mathrm{d}\rho}{\mathrm{d}s} = \frac{\partial\rho}{\partial t}\frac{\mathrm{d}t}{\mathrm{d}s} + \frac{\partial\rho}{\partial x}\frac{\mathrm{d}x}{\mathrm{d}s} = \frac{\partial\rho}{\partial t} \cdot 1 + \frac{\partial\rho}{\partial x} \cdot c(\rho(x, t)) = 0. \tag{7.29}$$

This means that ρ is constant along the field curves. Consequently, $c = c(\rho)$ is also constant along the curves. But this in turn implies that the curves are straight lines. These field lines are called *characteristic curves* or simply *characteristics*. Strictly speaking, the field lines are the projection in the (x, t)-plane of the real characteristics in (x, t, ρ)-space, but it is common also to call the projections of characteristics (in general, characteristic curves do not need to be straight lines).

We are now going to find the solution $\rho(x, t)$ for $-\infty < x < \infty$, $t \geq 0$, given that

$$\rho(x, 0) = f(x).$$

With the condition given at $t = 0$, this is an initial value problem, also called the *Cauchy problem* in this context.

However, if we are seeking the solution at a point (x_1, t_1), we first need to find the characteristic curve through the point. Since the characteristics are straight lines, they have equations

$$x = x_0 + c(\rho(x_0, 0)) t = x_0 + c(f(x_0)) t,$$

where $(x_0, 0)$ lies on the x-axis. Accordingly, we must first find an x_0 such that

$$x_1 = x_0 + c(\rho(x_0, 0)) t_1. \tag{7.30}$$

Solving (7.30) implies solving an implicit (and in general nonlinear) equation in order to find x_0. Once we know x_0,

$$\rho(x_1, t_1) = \rho(x_0, 0) = f(x_0), \tag{7.31}$$

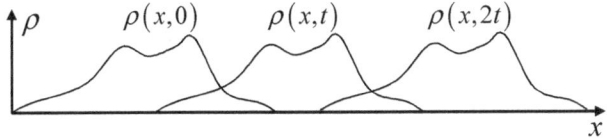

FIGURE 7.4
The solution of the simplest kinematic wave equation is a fixed function translated along the x-axis with constant speed.

since the value of ρ is constant along the characteristic. It is only possible to write the solution in explicit form in simple cases. Let us, as a very simple example, consider the equation

$$\frac{\partial \rho}{\partial t} + \frac{\partial \rho}{\partial x} = 0. \tag{7.32}$$

Here, $c(\rho) \equiv 1$ so that the characteristics are the lines in (x,t)-plane defined by $x = x_0 + t$. Thus, we get $\rho(x,t) = \rho(x_0,0) = f(x_0) = f(x-t)$. The variation in the density at $t = 0$ thus moves without changing the shape towards the right with speed 1 as illustrated in Figure 7.4.

7.3.1 The Riemann problem

Let us again consider the 1D conservation law in integral form,

$$\frac{\mathrm{d}}{\mathrm{d}t} \int_a^b \rho(x,t)\,\mathrm{d}x + j(\rho(b,t)) - j(\rho(a,t)) = 0, \tag{7.33}$$

with the differential formulation

$$\frac{\partial \rho}{\partial t} + c(\rho)\frac{\partial \rho}{\partial x} = 0, \ \ c(\rho) = \frac{\mathrm{d}j}{\mathrm{d}\rho}. \tag{7.34}$$

For this conservation law, there are three basic solutions that typically arise, and even if we can find this in most textbooks about Partial Differential Equations, for example the book by Whitham [41], we shall for completeness list them here as well. The three cases are solutions to the so-called *Riemann problem*, where we want to determine $\rho(x,t)$ for $-\infty < x < \infty$ and $0 \leq t$ when

$$\rho(x,0) = \begin{cases} \rho_1, & x < 0 \\ \rho_2, & x > 0 \end{cases}, \ \rho_1 \neq \rho_2.$$

The characteristics starting at x_0 on the x-axis, in this case, are given by

$$x = x_0 + c(\rho_1)t, \ \ x_0 < 0,$$
$$x = x_0 + c(\rho_2)t, \ \ x_0 > 0.$$

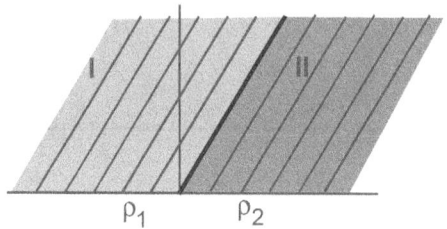

FIGURE 7.5
Contact discontinuity: the characteristics are parallel.

7.3.2   Contact discontinuity

If we have $c(\rho_2) = c(\rho_1)$, the characteristics will be parallel for both positive and negative x_0. The region $-\infty < x < \infty$, $0 \le t$, is now divided into two parts, I and II (see Figure 7.5) where the solutions ρ to the differential equation are, respectively, ρ_1 and ρ_2. The solution is called a *contact discontinuity*, since it is discontinuous along the contact of the two parts, i.e., the line $x = c(\rho_1)\,t = c(\rho_2)\,t$. Somewhat surprisingly, this solution does *not* need to be an acceptable solution for the conservation law. We shall see below that the conservation law is only satisfied if, in addition,

$$(\rho_1 - \rho_2)\,c(\rho_1) + j(\rho_2) - j(\rho_1) = 0 \qquad (7.35)$$

and this does not need be the case even if $c(\rho_1) = j'(\rho_2) = c(\rho_2)$. If this

extra condition is not met, the solution of the conservation law develops in a more complicated way. The basic situation is, however, when the condition is fulfilled.

7.3.3   Rarefaction wave

If $c(\rho_1) < c(\rho_2)$, the characteristics starting outside the origin have to go as shown in Figure 7.6. The solution in the regions I and III are thus ρ_1 and ρ_2, respectively.

If $c(\rho)$ is *monotonously increasing* when ρ goes from ρ_1 to ρ_2, the solution in region II becomes what is called an elementary *rarefaction wave*, *expansion wave*, or *expansion fan*. Here, all characteristics have to start at the origin. Therefore, the characteristics have all the equation $x = c(\rho)\,t$, and consequently, the solution for a point (x, t) in region II is given implicitly by

$$\rho(x, t) = c^{-1}(x/t) \qquad (7.36)$$

(The inverse function c^{-1} exists under the above assumption of monotonicity). We leave it to the reader to show that this solution really fulfills the conservation law.

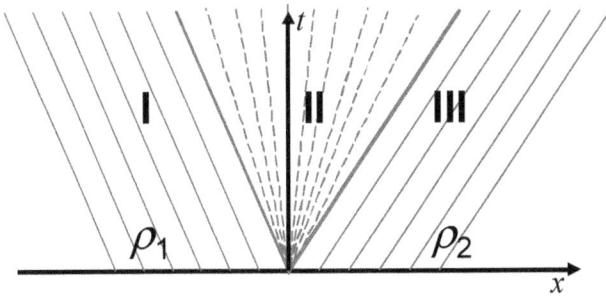

FIGURE 7.6
The rarefaction wave.

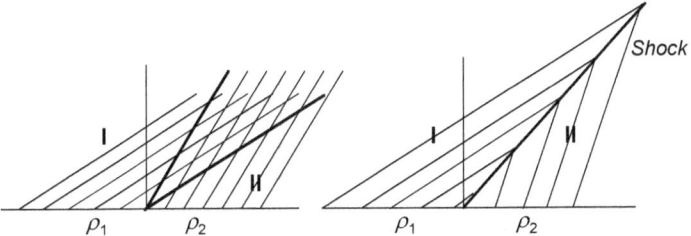

FIGURE 7.7
The situation when $c(\rho_1) > c(\rho_2)$ requires the introduction of a shock.

Consider the following simple example where $c(\rho) = \rho$, $\rho_1 = 0$, and $\rho_2 = 1$. Then the characteristics for $x_0 < 0$ are simply $x = x_0$, and $x = x_0 + t$ for $x_0 > 0$. In region II defined by $\{(x,t) ; 0 < x < t\}$ the solution is given by $x = \rho t$, i.e., $\rho = x/t$. Thus, the solution for $t > 0$ becomes

$$\rho(x,t) = \begin{cases} 1, & x < 0, \\ x/t, & 0 \le x \le t, \\ 0, & t < x. \end{cases} \tag{7.37}$$

Try some other possibilities for $c(\rho)$ and make sketches to see how things come out!

7.3.4 Shock solution

If we have the reverse situation from above, namely that $c(\rho_1) > c(\rho_2)$, the characteristics will cross, as illustrated in Figure 7.7 to the left. Although the solution outside the collision area can be found using the characteristic method, this is of little help in the area where the characteristics cross. In order to resolve the situation, we have to go back to the original conservation law and introduce a *discontinuity*, $x = s(t)$, called a *shock*, as illustrated in Figure 7.7 to the right.

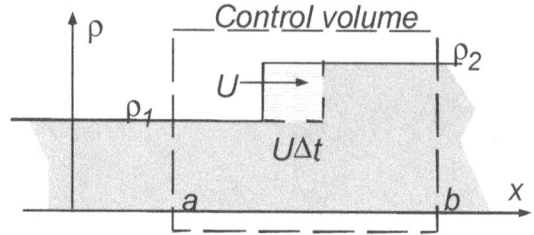

FIGURE 7.8
A control volume enclosing a discontinuity in the density. The change in the content during a time interval Δt is $(\rho_2 - \rho_1)\, U \Delta t$.

In order to determine the speed $U = \mathrm{d}s/\mathrm{d}t$ of the shock, we consider the interval $[a, b]$ such that it includes discontinuity. By calculating the change of the contents in $[a, b]$, as outlined in Figure 7.8, we obtain

$$\frac{\mathrm{d}}{\mathrm{d}t} \int_a^b \rho(x, t)\, \mathrm{d}t = \lim_{\Delta t \to 0} \frac{-(\rho_2 - \rho_1) \times U \Delta t}{\Delta t} = (\rho_1 - \rho_2)\, U. \tag{7.38}$$

The conservation law will now only be satisfied if

$$\frac{\mathrm{d}}{\mathrm{d}t} \int_a^b \rho(x, t)\, \mathrm{d}x + j\rho(b, t) - j(\rho(a, t)) = (\rho_1 - \rho_2)\, U + j(\rho_2) - j(\rho_1) = 0. \tag{7.39}$$

(recall the situation for the contact discontinuity). Thus we derive that the shock speed must be

$$U = \frac{j(\rho_2) - j(\rho_1)}{\rho_2 - \rho_1}, \tag{7.40}$$

and the solution for $t > 0$ is thus

$$\rho(x, t) = \begin{cases} \rho_1, & x < Ut \\ \rho_2, & x > Ut. \end{cases} \tag{7.41}$$

We shall finally show a simple example where the same differential equation may come from different conservation laws. Let the density and the flux depend on another function u such that

$$\rho(x, t) = \frac{1}{n} u(x, t)^n,$$

$$j(x, t) = \frac{1}{n+1} u(x, t)^{n+1}.$$

The differential formulation follows from the conservation law $\frac{\partial \rho}{\partial t} + \frac{\partial j}{\partial x} = 0$, which leads to

$$\frac{\partial \rho}{\partial t} + \frac{\partial j}{\partial x} = u^{n-1} \frac{\partial u}{\partial t} + u^n \frac{\partial u}{\partial x} = u^{n-1} \left(\frac{\partial u}{\partial t} + u \frac{\partial u}{\partial x} \right) = 0. \tag{7.42}$$

The differential equation for u is therefore essentially the same regardless of the value of n we had in the conservation law. The characteristics are the same in all cases, but the shock speeds are different for different n-s:

$$U = \frac{n}{n+1} \frac{u_2^{n+1} - u_1^{n+1}}{u_2^n - u_1^n}. \tag{7.43}$$

The example shows that if the solutions develop shocks, the shock can *not* be found from the differential equation alone. The position of the shock must be determined from the original conservation law. In some cases it is also necessary to bring in additional conditions in order to determine a physically acceptable solution (this is more thoroughly covered in courses in non-linear partial differential equations).

7.4 Exercises

1. (a) Define *density* and *flux* defined.
   (b) What is a *point source*, and how can it be described?
   (c) What is a *distributed source*, and how can it be expressed when the flux is known?
   (d) What is the flux for a substance that passively follows a fluid velocity field $\mathbf{v}(\mathbf{x},t)$? Show that the dimension of the expression is consistent with the definition of flux.
   (e) How do we derive the differential form of the integral form of the general conservation law?

2. A hemisphere with radius r has centre at the origin, and is bounded by $0 \leq z$. The flux field $\mathbf{j}$ is defined in the space as

$$\mathbf{j}(\mathbf{x},t) = (y \sin z)\,\hat{\imath}_x + xz^3\hat{\imath}_y + z\hat{\imath}_z,$$

where $\{\hat{\imath}_x, \hat{\imath}_y, \hat{\imath}_z\}$ are the unit vectors along the coordinate axes. Calculate the flow of material through the curved part of the surface of the hemisphere, i.e., $|\mathbf{x}| = r$, $z > 0$. *Hint:* The solution is simple.

8

Modelling of Road Traffic

Systematic studies of road traffic started about 70 years ago in the homeland of cars, the USA. During the 1950s mathematicians also began to get stuck in rush hour traffic, and more theoretical work appeared in mathematical journals. An article by M. J. Lighthill and G.B. Whitham in the Proceedings of the Royal Society entitled *On kinematic waves II. A theory of traffic flow on long, crowded roads*, from 1955 is one of the milestones in the development [29]. Mathematical modelling of road traffic is a relatively wide field, and there is much information on the Internet and in several textbooks.

The present material has been based on lectures and seminars in mathematical modelling at NTNU over several years. The report published by US Transportation Research Board, *Traffic Flow Theory* is, as of this writing, available free of charge from the Internet [17].

Research claims that between 20–30 percent of traffic jams on Norwegian roads would disappear if each motorist were driving more efficiently. Researchers at NTNU have in controlled trials managed to double the flux from 1800 to 3600 vehicles per hour just by adjusting the drivers' behaviour. Interestingly enough, it appears that sometimes the flux of cars may be larger when the traffic is kept at 60 km/h, compared to 90 km/h. On the ring-road around London, $M25$, the speed limits are constantly adjusted in order to optimize the traffic flow.

Traffic modelling can be approached from many sides and by applying many different mathematical and statistical tools. It is reasonable to think of models based on individual vehicles on a road where the speed is expressed as a function of road conditions and other vehicles nearby. The models may contain stochastic elements such as variations in the drivers' perception of what is a *safe speed*, a *safe distance to the vehicle in front*, and an *acceptable overtaking margin*. Such models quickly become analytically complicated, but are suitable for computer simulations. Queuing theory and other statistical models describing the randomness of real traffic are also widely used.

The article of Lighthill and Whitham suggests a continuum model for car traffic. Traffic flow is described in terms of density, flux, sources and sinks, which consequently leads to hyperbolic conservation laws. This theory is called the *kinematic theory* of road traffic. The models can be refined by including the drivers' ability to respond to changes in traffic density, and how quickly they can adjust according to road conditions. The material below is mainly taken from the books by Whitham and Haberman stated in the reference list.

DOI: 10.1201/9781003725206-8

8.1 Kinematic Theory

In kinematic traffic theory, traffic is modelled by means of a simple conservation law. The density $\rho(x,t)$ of cars on the road is expressed as the number of cars per unit length. The term must be considered somewhat pragmatic, as is often the case when we model a collection of discrete objects as a continuous medium. We consider ρ as a piecewise continuous function of position and time. Because of the car's finite size, it is reasonable to assume that

$$0 \le \rho \le \rho_{\max},$$

where $\rho_{\max}$ is the maximum density as calculated from the cars' average length.

The car velocity v is assumed to be a function of car density, $v(\rho)$, so that $v(0) = v_{\max}$ and $v(\rho_{\max}) = 0$. Thus, v decreases as ρ increases. A clear-cut relation between the cars' velocity and density may only be reasonable on one-lane roads, but are also used for multi-lane highways, where the car velocities vary both individually and from lane to lane. In this case, one interprets v as the average speed, and measurements indicate this is a reasonable assumption, at least for parts of the interval between 0 and $\rho_{\max}$.

If we assume that the speed of the cars is a function of ρ, $v(\rho)$, a small argument gives us that the flux of cars, i.e., the number of cars passing a given point on the road per time unit, can be expressed as

$$J = \rho v(\rho).$$

In the traffic literature flux is often designated with the symbols k, or F, and the graph of J as a function of ρ is called the *fundamental diagram*. Normally, $J(\rho)$ tends to 0 when ρ approaches 0 or $\rho_{\max}$, and is a concave function with a maximum value somewhere between 0 and $\rho_{\max}$. The conservation law becomes as before

$$\frac{\mathrm{d}}{\mathrm{d}t} \int_a^b \rho(x,t)\mathrm{d}x + J(b,t) - J(a,t) = \int_a^b q(x,t)\mathrm{d}x, \qquad (8.1)$$

where the source term expresses cars entering or leaving the road. The differential formulation leads to a first-order hyperbolic equation, and since J is only a function of ρ, we can write the equation

$$\frac{\partial \rho}{\partial t} + \frac{\partial J}{\partial x} = \frac{\partial \rho}{\partial t} + \frac{\mathrm{d}J}{\mathrm{d}\rho}\frac{\partial \rho}{\partial x} = q. \qquad (8.2)$$

If $q = 0$, the characteristics will be straight lines with slope (kinematic velocity) equal to $c(\rho) = \frac{\mathrm{d}J}{\mathrm{d}\rho}$.

By applying the equations (8.1) and (8.2), we can examine what is happening around a traffic light crossing when we have a varying density of traffic,

how the individual cars are moving, etc. The car's own motion is determined
by a differential equation

$$\frac{\mathrm{d}x}{\mathrm{d}t} = v(\rho(x, t)).$$

Usually, $c(\rho)$ will be a decreasing function of ρ. This will typically lead to
situations developing shocks if the traffic is moving in the positive x-direction
and ρ increases with x. From the conservation law, a shock $x = s(t)$ will have
to satisfy

$$\frac{\mathrm{d}s}{\mathrm{d}t} = \frac{J(s+, t) - J(s-, t)}{\rho(s+, t) - \rho(s-, t)},$$

Changes in traffic conditions can be incorporated in several ways. If the
road has a narrowing, e.g., goes from two lane to one, it is reasonable that $\rho_{\max}$
is reduced, whereas $v_{\max}$ remains the same. This changes the fundamental
diagram. At the start of the narrowing, the flux has to be continuous, whereas
ρ will have a discontinuity.

If the road is slippery and the visibility is poor due to rain or fog, then $v_{\max}$
will decrease whereas $\rho_{\max}$ remains unchanged. This changes the fundamental
diagram in a different way.

A very common kinematic model that is reasonably easy to work with
analytically, is to assume that v is a decreasing, linear function of ρ. After
scaling, we obtain the equations

$$
\begin{aligned}
v(\rho) &= 1 - \rho, \\
J(\rho) &= \rho(1 - \rho), \\
c(\rho) &= 1 - 2\rho, \\
\frac{\partial \rho}{\partial t} + (1 - 2\rho)\frac{\partial \rho}{\partial x} &= 0.
\end{aligned}
\tag{8.3}
$$

Figure 8.1 shows how the car velocity, the flux, and the kinematic velocity
change for this model. Note that $c(\rho) = \frac{\mathrm{d}}{\mathrm{d}\rho}(\rho v) = v + \rho v'(\rho)$, and therefore
$c(\rho) \neq v(\rho)$ when $\rho \neq 0$. Below we shall use this model to analyze some
simple situations.

8.1.1   Traffic lights

Assume that at $x = 0$, there has been a red light for the cars for $t < 0$. To
the left of the light ($x < 0$), there is a dense queue of cars, $\rho = 1$, while to the
right ($x > 0$), there are no cars and $\rho = 0$. In all such problems, it is useful to
outline a so-called x/t-chart that describes the conditions, in particular how
the characteristics behave, as shown in Figure 8.2. When $t > 0$, we have three
regions. To the left is an area where $\rho = 1$, to the right a region where $\rho = 0$,
while in the middle there is an expansion wave with characteristics starting
at the origin,

$$x = c(\rho)t = (1 - 2\rho)t.$$

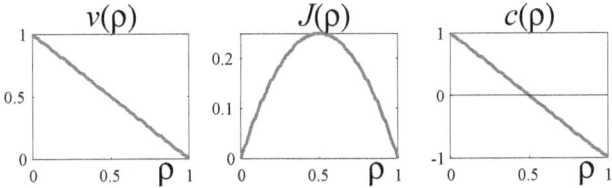

FIGURE 8.1
Speed, flux and kinematic velocity as a function of car density for the standard model.

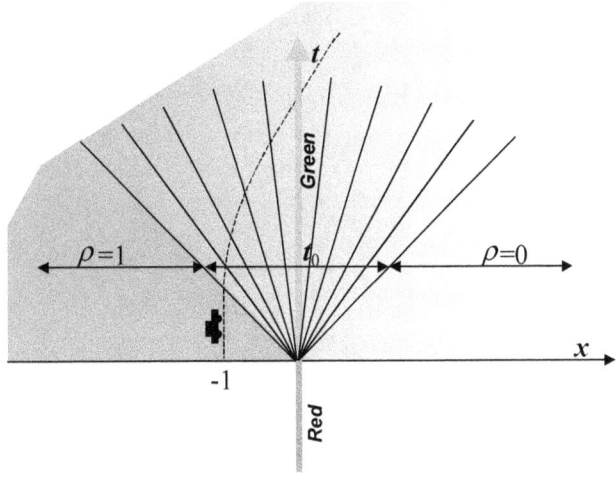

FIGURE 8.2
Junction at $x = 0$, where the light changes from red to green.

Within the central area $\rho = \frac{1}{2}\left(1 - \frac{x}{t}\right)$, and the complete solution becomes for $t = t_0$

$$\rho(x, t_0) = \begin{cases} 1, & x \leq t_0, \\ \frac{1}{2}\left(1 - \frac{x}{t_0}\right) & -t_0 < x < t_0, \\ 0 & x \geq t_0. \end{cases}$$

Suppose you are seated in a car at $x = -1$ for $t = 0$. What will be your own motion for $t > 0$? From Figure 8.2, we observe that you will start driving at $t = 1$, and then your own path, $y(t)$, will be controlled by the differential equation

$$\frac{dy}{dt} = v(\rho) = 1 - \frac{1}{2}\left(1 - \frac{y}{t}\right)$$
$$= \frac{1}{2}\left(1 + \frac{y}{t}\right), \quad y(1) = -1.$$

The equation is thus $2t\dot{y} - y = t$, with general solution $y(t) = At^{1/2} + t$. Since $y(1) = -1$, $y(t) = t - 2t^{1/2}$. It is worth noting that the distance to the front car at $x = t$ becomes longer as time passes.

The situation we have analyzed resembles what you encounter in a big running event: if you have ambitions of fighting in the lead, it pays to have a position as close as possible to the head of the queue before the start.

8.1.2   Traffic clogging up

Assume that there is a line of traffic on the road where the density at $t = 0$ has the form

$$\rho(x,0) = \begin{cases} \rho_1, & x \le a, \\ \rho_1 + \frac{x-a}{b-a}(\rho_2 - \rho_1) & a < x < b, \\ \rho_2 & b \le x, \end{cases}$$

where $\rho_1 < \rho_2$. Between $x = a$ and $x = b$, the car density increases linearly from ρ_1 to ρ_2. The characteristics are given by $x = x_0 + (1 - 2\rho)t$, and for characteristics between a and b this amounts to

$$x = x_0 + t - 2\left(\rho_1 + \frac{x_0 - a}{b - a}(\rho_2 - \rho_1)\right)t$$

$$= x_0 + (1 - 2\rho_1)t - 2\frac{x_0 - a}{b - a}(\rho_2 - \rho_1)t.$$

By inserting the time $t_s = \frac{1}{2}\frac{b-a}{\rho_2 - \rho_1}$, we see that x_0 vanishes. This means that all characteristics starting from the interval $[a, b]$ meet in the point (x_s, t_s),

$$x_s = a + (1 - 2\rho_1)\frac{1}{2}\frac{b - a}{\rho_2 - \rho_1} = b + (1 - 2\rho_2)\frac{1}{2}\frac{b - a}{\rho_2 - \rho_1}.$$

The situation is sketched in Figure 8.3. For $t > t_s$, we get a jump in density, a *shock*. The speed of the shock has to be determined from the shock condition, as discussed above,

$$U = \frac{J(\rho_2) - J(\rho_1)}{\rho_2 - \rho_1}$$

$$= \frac{\rho_2(1 - \rho_2) - \rho_1(1 - \rho_1)}{\rho_2 - \rho_1}$$

$$= 1 - \rho_1 - \rho_2.$$

One may wonder what is happening around such a shock, and in practice, the cars will try to avoid colliding. However, when driving through the shock, the car velocity has a discontinuity, and it is limited how fast it is possible to react!

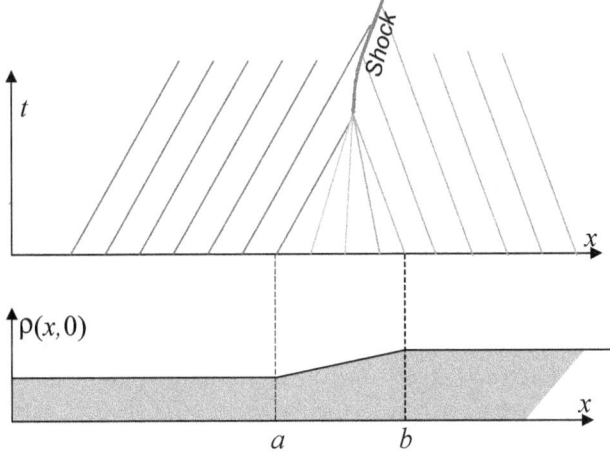

FIGURE 8.3
The situation with an increase in the car density.

8.1.3 When is the first shock formed?

The shock is formed when two characteristics collide. Let us look at two characteristics starting at x_0 and $x_0 + \Delta x$, respectively, and meeting at (x_s, t_s):

$$x_0 + c(x_0)t_s = x_0 + \Delta x + c(x_0 + \Delta x)t_s.$$

Thus,

$$t_s = -\frac{1}{(c(x_0 + \Delta x) - c(x_0))/\Delta x}.$$

If we let $\Delta x \to 0$, the limit is

$$t_s = -\frac{1}{\frac{dc}{dx}\big|_{x_0}}.$$

For a shock to form for $t > 0$, there has to be some x_0 where $\frac{dc}{dx}\big|_{x_0} < 0$, and the first time this happens is

$$\min t_s = -\frac{1}{\min_x \frac{dc}{dx}}.$$

8.1.4 Narrowing the road

On a road with two lanes for $x < 1$, one of the lanes is closed for $x > 1$, so that the maximum car density for $x > 1$ is only half of the original. Table 8.1 shows how the flux varies with the density of each of the two parts of the road. There is no storage for cars at $x = 1$. Therefore, the flux around

TABLE 8.1

The conditions surrounding a narrowing of the road.

Region	ρ	$v(\rho)$	$J(\rho)$	$c(\rho)$
$x < 1$	$[0\ 1]$	$1 - \rho$	$\rho(1 - \rho)$	$1 - 2\rho$
$x > 1$	$[0\ \frac{1}{2}]$	$1 - 2\rho$	$\rho(1 - 2\rho)$	$1 - 4\rho$

$x = 1$ must be continuous,

$$J\left(\rho(1-, t)\right) = J\left(\rho(1+, t)\right).$$

This means that the density is discontinuous (if different from 0). If the number of vehicles passing $x = 1$ is as large as possible, the flux is $J = 1/8$, and densities immediately to the left and right of $x = 1$ are given by

$$J\left(\rho^-\right) = \rho^-\left(1 - \rho^-\right) = J\left(\rho^+\right) = \rho^+\left(1 - 2\rho^+\right).$$

Thus,

$$\rho^+ = 1/4,$$

whereas there are two possibilities for ρ^-:

$$\rho^- = \frac{1}{2}\left(1 \pm \frac{\sqrt{2}}{2}\right).$$

To see how both possibilities can occur, we connect this situation with a traffic light at $x = 0$, as in the first example. The situation is illustrated in Figure 8.4. We get a shock at $x = 1$ until the density ρ on the left side reaches the value $\rho_1^- = \frac{1}{2}\left(1 - 2^{-1/2}\right)$. Then we get a sudden jump in the density up to $\rho_2^- = \frac{1}{2}\left(1 + 2^{-1/2}\right)$. Before the narrowing, we get a queue where the density is ρ_2^-. At the end of this queue, another shock is formed. We leave it to readers to consider what it is like to drive through such a situation.

8.1.5   Research project: *A green wave in infinity street?*

In a long straight street, the pedestrian crossings are organized with traffic lights. If the traffic lights are uncoordinated, the cars will need to drive and stop at uneven intervals, and the resulting average flux of cars may be quite low. However, sometimes we hear that it is possible to arrange the lights in a so-called green wave, so that the cars may "surf" through the street without having to stop. Is it really possible to streamline the traffic by using the green waves, and what is the maximum possible average flux? In this study, we will consider an idealized situation of this problem.

At *Infinity Street* pedestrian crossings are located at a constant distance L. There are no side streets with opportunities for the cars to leave or enter the street. The crossings have all traffic lights with a cycle of length S. This

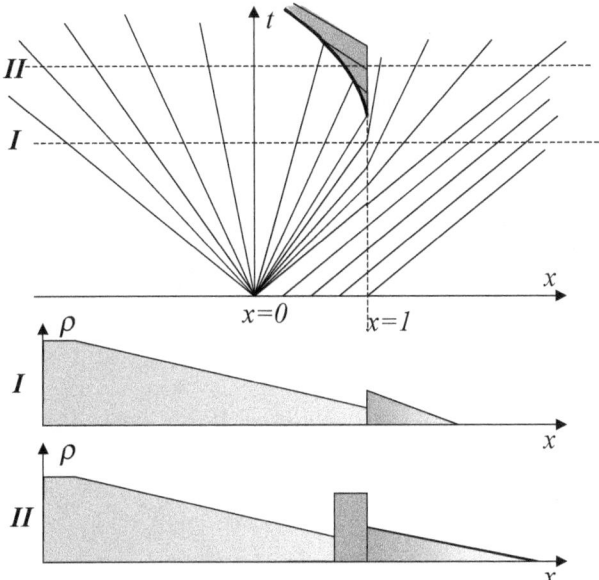

FIGURE 8.4
Narrowing of the road after a traffic light.

means that it is *red* in the period $[0, r^*)$, then it is *green* in the period $[r^*, S)$. The cycle repeats itself continuously. To facilitate the driving, the light cycles may be displaced in time in relation to each other, so that the cycle of crossing $k + 1$ starts some time before or after the cycle of crossing k. The maximum vehicle density is $\rho_{\max}$, and the velocity is $v_{\max}$. Traffic follows the simple kinematic model considered above. By scaling x and ρ the usual way, and using the time scale $L/v_{\max}$ we obtain (8.3).

The length of the red period is denoted by r and the green period by g. Here $r = 1$ corresponds to the shortest time it takes to drive between two traffic lights separated by a length 1, and the cycle length is $r + g$. The traffic goes around the clock, and the problem is to maximize the average flux, $\bar{J}$. If the red period starts in $x = 0$ at $t = 0$, the average flux will be

$$\bar{J} = \frac{1}{r+g} \int_0^{r+g} J(0,t)\mathrm{dt} = \frac{1}{r+g} \int_r^{r+g} \rho(0,t)[1 - \rho(0,t)]\mathrm{dt}.$$

It is clear that the theoretical maximum average flux will be

$$\bar{J}_{\max} = \frac{g}{r+g} J_{\max} = \frac{g}{r+g} \frac{1}{4},$$

since the maximum number of cars passes through the crossing during the green period, and anything better than that is impossible.

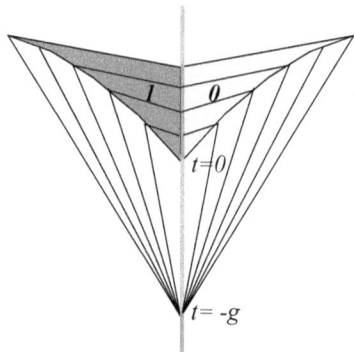

FIGURE 8.5
The expansion wave from $(0, -g)$ collides with the shocks that start at $t = 0$. The characteristics in the regions where $\rho = 0$ and 1 have speed 1 and -1, respectively.

We shall first analyse a simpler situation with one traffic light. Until time $t = -g$ there has been red light at $x = 0$. To the left of $x = 0$ there is a queue of cars. The light is green up to $t = 0$, where it again changes to red. From the point $(0, -g)$ there is now an expansion wave which collides with the shock from $(0, 0)$ in both positive and negative x-directions, as shown in Figure 8.5. The shock speed in this case is

$$U = \frac{J_2 - J_1}{\rho_2 - \rho_1} = 1 - (\rho_1 + \rho_2).$$

The density of the two characteristics that are symmetrical about the t-axis are $\rho^- = \frac{1}{2} - \rho$ and $\rho^+ = \frac{1}{2} + \sigma$, respectively. Thus, the two shock speeds are the same with the opposite sign:

$$U^+ = 1 - (0 + \frac{1}{2} - \sigma) = \frac{1}{2} + \sigma,$$
$$U^- = 1 - (1 + \frac{1}{2} + \sigma) = -\left(\frac{1}{2} + \sigma\right).$$

Consequently, the two shock curves are also symmetrical about the origin. We can calculate the shock curve x_c from

$$\frac{dx_s}{dt} = U^+(x_s, t) = 1 - \rho^+$$
$$= 1 - c^{-1}\left(\frac{x_s}{t + g}\right)$$
$$= 1 - \frac{1}{2}\left(1 - \frac{x_s}{t + g}\right),$$

or
$$2(t + g)\dot{x}_s = (t + g) + x_s.$$

The solution is
$$x_s(t) = t + g - g^{1/2}(t + g)^{1/2}.$$

The shocks follow this curve until they meet the characteristics that have started in $t = r$, in other words, immediately after the red period from $t = 0$ to r is over. The areas with $\rho = 0$ and $\rho = 1$ are therefore also *symmetric* about the y-axis.

Let us now look at a situation where the red period from $x = 0$ has lasted so long that the areas in Figure 8.5 reach out to $x = \pm\frac{1}{2}$. We find out where $(\pm\frac{1}{2}, t_0)$ is by observing that

$$\frac{1}{2} = 0 + c(0)(t_0 - r),$$

that is,

$$t_0 = r + \frac{1}{2}\frac{1}{c(\rho = 0)} = r + \frac{1}{2}.$$

Then we put this into the equation for the shock:

$$\frac{1}{2} = (t_0 + g) - g^{1/2}(t_0 + g)^{1/2}$$
$$= \left((r + \frac{1}{2}) + g \right) - g^{1/2} \left((r + \frac{1}{2}) + g \right)^{1/2},$$

or

$$g = \frac{r^2}{\frac{1}{2} - r}.$$

The solution requires that $r < 1/2$, and otherwise that $g = r$ for $r = 1/4$. For these combinations of g and r, we can now construct complete solutions for *Infinity Street* that for any given ratio $r/g \in \mathbb{R}^+$ provide maximum throughput. The construction is best described on a figure (see Figure 8.6). Note that we have got a motionless shock for half integers (the shock speed is 0, since the sum of the densities on both sides is 1 at any time). If we want r/g to be large, this gives very short cycles. Even the maximum red period, $r = 1/2$, which corresponds to the time it takes to cover half the distance between two traffic lights at maximum speed, seems to be rather short for practical purposes.

It appears that the symmetric structure may be generalized:

- At shorter distances between the crossings, the "leaves" are cut and meet in the middle.

- For larger distances the "leaves" are extended with two shocks that also meet in the middle.

Is it possible to have a solution that looks like the one in Figure 8.7?

FIGURE 8.6
Sketch of the complete solution. Note that we get a stationary shock at $x = 1/2$.

8.2   Generalizations of the Kinetic Theory

In practice, drivers try to compensate for changes in traffic density by adjusting the speed according to surrounding conditions. To avoid the development of shocks and thus extremely rapid changes in density, they will tend to slow down somewhat more than the relationship $v = v(\rho)$ suggests, and therefore avoid strong gradients in density build up. Whitham models this by assuming that the drivers adjust the speed as the

$$v = v_k(\rho) - \kappa\frac{\partial\rho/\partial x}{\rho}, \qquad (8.4)$$

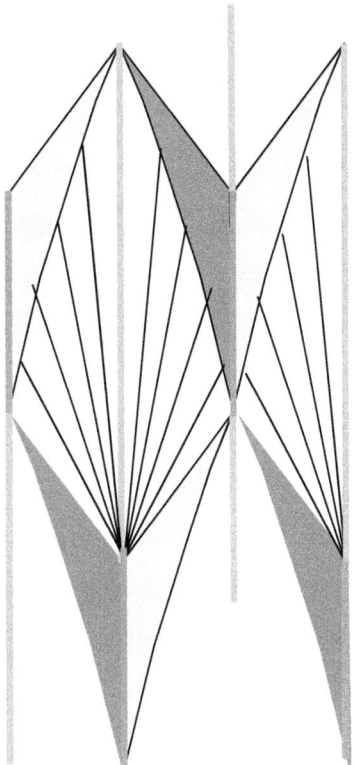

FIGURE 8.7
Is this a possible solution?

where v_k is the speed according to the kinematic theory above. This seems to be a reasonable model since $v = v_k(\rho)$ if ρ is constant, whereas v is less than v_k if ρ increases in the driving direction, and the opposite when ρ decreases. The best part of the model, and probably the reason why Whitham selected the particular form in (8.4), is that it leads to a famous 2nd order parabolic equation for ρ (since $J = \rho v = \rho v_k(\rho) - \kappa \rho_x$):

$$\frac{\partial \rho}{\partial t} + c(\rho)\frac{\partial \rho}{\partial x} - \kappa \frac{\partial^2 \rho}{\partial x^2} = 0, \ c(\rho) = \frac{\mathrm{d}(\rho v_k)}{\mathrm{d}\rho}.$$

For the special case where $c(\rho) = 1 - 2\rho$ we may use c instead of ρ as the dependent variable, leading to the equation

$$\frac{\partial c}{\partial t} + c\frac{\partial c}{\partial x} - \kappa \frac{\partial^2 c}{\partial x^2} = 0, \tag{8.5}$$

called *Burgers equation*. Burgers equation is one of the most studied non-linear partial differential equations and is extensively described in, e.g. [41], Ch. 4.

It turns out that Burgers equation has solutions in the form of "migrating fronts":

$$c(x,t) = C(s), \ s = x - x_0 - at,$$
$$\lim_{s \to -\infty} C(s) = c_1, \ \lim_{s \to \infty} C(s) = c_2.$$

The front thus has a constant value on the curves $x = x_1 + at$, and well-defined limits as x tend to $\pm\infty$. We find such a solution by entering C in the equation:

$$-aC' + CC' - \kappa C'' = 0.$$

By one integration we obtain

$$(C - a)^2 = 2\kappa C' + A,$$

where A is another integration constant. Since we expect that the derivative also goes to 0 when $x \to \pm\infty$, we see that

$$a = \frac{c_1 + c_2}{2},$$

$$A = \left(\frac{c_1 - c_2}{2}\right)^2.$$

We now insert this and separate the variables:

$$-\frac{ds}{2\kappa} = \frac{dC}{\left(\frac{c_2 - c_1}{2}\right)^2 - (C - a)^2}.$$

This gives

$$-\frac{s - s_0}{2\kappa} = \frac{2}{c_2 - c_1} \operatorname{arctanh}\left(2\frac{C - a}{c_2 - c_1}\right),$$

which may be turned around to

$$C(s) = \frac{c_1 + c_2}{2} - \frac{c_2 - c_1}{2} \tanh\left(\frac{c_2 - c_1}{4\kappa}(s - s_0)\right)$$

$$= c_1 + (c_2 - c_1)\left[1 + \exp\left(\frac{c_2 - c_1}{2\kappa}(s - s_0)\right)\right]^{-1}.$$

The function is a *sigmoid* (recall the equation for logistic growth). If $c_2 < c_1$, then $C(s)$ is a monotone decreasing front moving at velocity

$$U = a = \frac{c_1 + c_2}{2}$$

from left to right. Note that since $c = 1 - 2\rho$ in this model, $\rho(x,t)$ will be a monotone *increasing* front moving with the same velocity. Moreover,

$$\lim_{x \to -\infty} \rho(x,t) = \rho_1 = \frac{1 - c_1}{2},$$

$$\lim_{x \to \infty} \rho(x,t) = \rho_2 = \frac{1 - c_2}{2},$$

$$U = \frac{c_1 + c_2}{2} = 1 - (\rho_1 + \rho_2).$$

The transition between $\rho = \rho_1$ and $\rho = \rho_2$ becomes sharper the smaller κ is, but U is independent of κ. We see further that U is just the shock speed for the equation

$$\rho_t + \rho\rho_x = 0$$

for the initial conditions $\rho(x,0) = \rho_1$ for $x < 0$, and $\rho(x,0) = \rho_2$ for $x > 0$, and $\rho_1 < \rho_2$. The behaviour in the limit $\kappa \to 0$ is therefore reasonable.

Whitham takes the theory a step further by studying what happens when the cars need some time to adjust to the proper speed. A driver has some finite reaction time. If we look at the speed of a particular car, v_B, in a situation where traffic density changes, the car velocity could probably follow the equation

$$\frac{dv_B}{dt} = \frac{1}{\tau}(V - v_B),$$

where V is the ideal speed the drivers want to follow, as given by (8.6). This is a well-known equation from control theory. If $v_B = v_0$ at time $t = 0$ and V is a given function of t, it is possible to express v_B as

$$v_B(t) = v_0 e^{-t/\tau} + \int_0^t \exp(\frac{t' - t}{\tau})V(t')dt'.$$

If $V = V_0$ is constant for $t > 0$, the solution becomes

$$v_B(t) = (v_0 - V_0)e^{-t/\tau} + V_0.$$

The effect of the initial value will die out with time constant τ, and if V varies slowly (that is, the time scale for variations in V is longer than τ), v_B will tend to follow V.

We want to formulate (8.6) for the velocity v as a function of both t and x. The car follows a path $x(t)$, so that $v_B(t) = v(x(t), t)$. This requires what in mechanics is called the convective derivative,

$$\frac{dv_B}{dt} = \frac{d}{dt}v(x(t), t) = \frac{\partial v}{\partial x}\frac{dx}{dt} + \frac{\partial v}{\partial t} = \frac{\partial v}{\partial x}v + \frac{\partial v}{\partial t}.$$

Altogether, this gives *two* coupled equations for the motion

$$\frac{\partial \rho}{\partial t} + \frac{\partial(\rho v)}{\partial x} = 0,$$

$$\frac{\partial v}{\partial t} + v\frac{\partial v}{\partial x} = \frac{1}{\tau}\left(\left(v_k(\rho) - \kappa\frac{\partial \rho/\partial x}{\rho}\right) - v\right).$$

The equations are called a *hyperbolic system*, and we shall not go into the solution of them in full detail, but we immediately see that the system has *equilibrium solutions*:

$$\rho = \rho_0,$$

$$v_0 = v_k(\rho_0).$$

The way of investigating whether the equilibria are stable or unstable is to write

$$\rho = \rho_0 + r,$$
$$v = v_0 + w,$$

where $r(x,t)$ and $w(x,t)$ are small perturbations. After some arithmetic, we arrive at the following system of equations, where we have inserted $c_0 = \frac{d}{d\rho}(v_k \rho)|_{\rho=\rho_0} = v_k'(\rho_0)\rho_0 + v_0$:

$$r_t + v_0 r_x + \rho_0 w_x = 0,$$
$$w_t + v_0 w_x = -\frac{1}{\tau}\left(w - \frac{c_0 - v_0}{\rho_0}r + \frac{\kappa}{\rho_0}r_x\right).$$

The equations may be combined into a single linear 2nd order equation containing only r:

$$r_t + c_0 r_x = \kappa r_{xx} - \tau\left(\frac{\partial}{\partial t} + v_0\frac{\partial}{\partial x}\right)^2 r.$$

We may recognize the first part,

$$r_t + c_0 r_x = \kappa r_{xx},$$

as a linear convection/diffusion equation, which is known to have solutions quickly dying out with time. However, the last term, $-\tau\left(\frac{\partial}{\partial t} + v_0\frac{\partial}{\partial x}\right)^2 r$, may create problems for us if it "dominates" the diffusion term κr_{xx}. The standard method for studying such linear equations is to examine *Fourier components*:

$$r(x,t) = ae^{i(kx-\omega t)}.$$

The Fourier component is a travelling wave with wave number k and frequency ω. If this is inserted into the equation, we get a so-called *dispersion relation* linking k and ω:

$$\tau(\omega - v_0 k)^2 + i(\omega - c_0 k) - \kappa k^2 = 0.$$

Since the wave number $k = 2\pi/\lambda$, where λ is the wavelength and k is a real number. On the other hand, the frequency will generally be complex. Moreover, since

$$ae^{i(kx-\omega t)} = ae^{(\mathrm{Im}\,\omega)t}e^{i(kx-(\mathrm{Re}\,\omega)t)},$$

we see that if the imaginary part of ω is greater than 0, the amplitude of the Fourier component will grow exponentially in time, while it decreases exponentially if $\mathrm{Im}\,\omega < 0$. It is possible to show that

$$\mathrm{Im}\,\sqrt{\frac{\kappa}{\tau}} < c_0 < v_0 + \sqrt{\frac{\kappa}{\tau}}.$$

This tells us that v_0 and c_0 cannot be too different. For all Fourier components to die out we must have

$$|v_0 - c_0| < \sqrt{\frac{\kappa}{\tau}}.$$

If we introduce $v_k(\rho) = 1 - \rho$, we see that $v_0 = 1 - \rho_0$ and $c_0 = 1 - 2\rho_0$ (in dimensionless variables). Therefore, the perturbations above die out only when

$$\rho < \sqrt{\frac{\kappa}{\tau}}.$$

This result, which does not seem to be mentioned in Whitham, is interesting in the light of observations that have been reported. The traffic seems to follow a fundamental diagram from $\rho = 0$ and up to a certain ρ_c which is considerably less than $\rho_{\max}$. For larger densities the well-defined behaviour breaks down, and J is significantly smaller than the fundamental diagram would indicate. It could therefore be interesting to know whether this is due to instabilities of the type we found here, but the problem is to find realistic numerical values to insert for τ and κ.

8.3 Individual Car Models

The theory in this section is mostly obtained from the book by Haberman, Sec. 64 [22].

Individual car models deal with individual cars on the road. With one lane, and no possibility of passing, we can assume that each car adjusts its own speed relative to the speed of the car ahead (assuming it is so close that the driver can see it). Let us denote the position of the vehicle number n in the queue by $x_n(t)$. Then we have the model

$$\frac{\mathrm{d}^2 x_n}{\mathrm{d}t^2} = -\lambda \left(\frac{\mathrm{d}x_n}{\mathrm{d}t} - \frac{\mathrm{d}x_{n-1}}{\mathrm{d}t} \right). \tag{8.6}$$

If car number n has higher speed than the vehicle in front (number $n - 1$), car number n will brake. Actually, there will be some delay in the reaction of drivers, so we should write

$$\frac{\mathrm{d}^2 x_n(t+T)}{\mathrm{d}t^2} = -\lambda \left(\frac{\mathrm{d}x_n(t)}{\mathrm{d}t} - \frac{\mathrm{d}x_{n-1}(t)}{\mathrm{d}t} \right), \tag{8.7}$$

where T expresses the drivers' response time. The equation may be integrated once:

$$\frac{\mathrm{d}x_n(t+T)}{\mathrm{d}t} = -\lambda \left(x_n(t) - x_{n-1}(t) \right) + d_n. \tag{8.8}$$

In a uniform situation where all cars have the same speed and same distance between them, the density will be

$$\rho = \frac{1}{x_{n-1} - x_n}.$$

In that case, the cars' velocity is thus

$$v = \frac{\mathrm{d}x_n}{\mathrm{d}t} = -\frac{\lambda}{-\rho} + d.$$

The constant d is chosen so that $v = 0$ for $\rho = \rho_{\max}$, and this leads to the following expression for v and the flux J:

$$v = \lambda \left(\frac{1}{\rho} - \frac{1}{\rho_{\max}} \right),$$

$$J = v\rho = \lambda\rho \left(\frac{1}{\rho} - \frac{1}{\rho_{\max}} \right).$$

The model is called the *California model*. The model is not very realistic for small densities since $v \to \infty$ when $\rho \to 0$. If there are very few cars on the road, there will be no cars in sight most of the time, and then it is reasonable to move at maximum speed. A modified model would be to set

$$v = \min \left(v_{\max}, \lambda \left(\frac{1}{\rho} - \frac{1}{\rho_{\max}} \right) \right).$$

It is also possible to embed the drivers' sensitivity for changes by assuming that λ varies with the distance to the car in front. We could for example assume that

$$\lambda = \frac{a}{x_{n-1} - x_n},$$

which means that the sensitivity disappears when the distance is great. The Eq. (8.6) then modifies to

$$\frac{\mathrm{d}^2 x_n}{\mathrm{d}t^2} = -\frac{a}{x_{n-1} - x_n} \left(\frac{\mathrm{d}x_n}{\mathrm{d}t} - \frac{\mathrm{d}x_{n-1}}{\mathrm{d}t} \right),$$

which can still be integrated analytically to

$$\frac{\mathrm{d}x_n(t + T)}{\mathrm{d}t} = -a \ln |x_n(t) - x_{n-1}(t)| + d_n.$$

An argument similar to the above yields

$$v = a \ln \frac{\rho_{\max}}{\rho},$$

$$J = \rho a \ln \frac{\rho_{\max}}{\rho}.$$

This still causes $v \to \infty$ when $\rho \to 0$, but here $J \to 0$ for both $\rho = 0$ and $\rho = \rho_{\text{max}}$. The model is called the *Greenberg model*, and, according to [41], it fits the traffic in the Lincoln Tunnel from Manhattan to New Jersey, when

$$a = 17.2 \text{ mph},$$
$$\rho_{\text{max}} = 228 \text{ cars/mile}.$$

The model is still not entirely satisfactory, but the same idea could be taken further, for example, assuming that

$$\lambda = \frac{\tilde{a}}{(x_{n-1} - x_n)^2}.$$

We then get a linear relationship between v and ρ. Measurements of flux as a function of density suggest a finite derivative at $\rho = 0$, in line with v going to a finite value when $\rho \to 0$.

8.3.1 Instabilities in a queue

Driving in a queue, we may have experienced that it is uncomfortable to be behind drivers who keep an irregular speed, and this may also be analyzed with individual car models. With a finite reaction time, the equations are no longer pure differential equations, but called *delay equations*. Consider two cars and assume that the speed v_1 of the car in front varies periodically. It turns out to be convenient to work with complex solutions, but the result using physical real periodic solutions will be the same. We therefore assume that

$$v_1(t) = 1 + ae^{i\omega t},$$

where the amplitude a is much smaller than 1. Assume that we have reached a stationary situation where the following car has a similar variation in the velocity,

$$v_2(t) = 1 + be^{i\omega t}.$$

We put this into (8.8) and get

$$bi\omega e^{i\omega T} = -\lambda(b - a),$$

or

$$b = \frac{1}{1 + \frac{i\omega}{\lambda}e^{i\omega T}} a.$$

Thus

$$\frac{|b|}{|a|} = \left| \frac{1}{1 + \frac{i\omega}{\lambda}e^{i\omega T}} \right| = \frac{1}{\sqrt{\left(1 - 2\frac{\omega}{\lambda}\sin\omega T + \left(\frac{\omega}{\lambda}\right)^2\right)}}.$$

The amplitude of b will be greater than the amplitude of a if the denominator is less than 1. This occurs when

$$\sin(\omega T) > \frac{\omega}{2\lambda},$$

which can be expressed as

$$\frac{\sin \omega T}{\omega T} > \frac{1}{2\lambda T}.$$

Since $\sin x / x \leq 1$, there is no danger as long as $\lambda T < 1/2$, but if this is not the case, there are low frequencies where the amplitude for the second car is larger than for the car in front. With several cars in the line, a further magnification will occur for cars further back. Good drivers will notice this and try to dampen the fluctuations in the speed. A similar analysis could also be considered for a sudden braking of the first car.

9

Conservation Laws of Mechanics

It turns out that the physics of the continuum matter surrounding us (for example, solid material, liquids, and gases) can be described in a compact way using the framework above, and this is acknowledged in most recent books about continuum and fluid mechanics [35].

However, the laws of physics are basically laws for a given collection of matter. For example, Newton's laws are laws for one or more "mass points". In the same way, a thermodynamic system, as we consider it when formulating the first law of thermodynamics, consists of a fixed collection of molecules. In continua like liquids and gases, where the material is moving, we are mostly interested in formulating the laws for a fixed region of space, that is, a *control volume*. The control volume will therefore contain different mass particles at different times.

In mechanics, a continuous medium in motion can be described in the *Eulerian* way by considering the velocity $\mathbf{v} = \mathbf{v}(\mathbf{x}, t)$ at each point $\mathbf{x}$, or we may use a *Lagrangian* description in which we follow the mass particles as time goes by, $\mathbf{x} = \mathbf{x}(t, \mathbf{a})$, $\mathbf{a} = \mathbf{x}(0, \mathbf{a})$.

A *material region* $R(t)$ is a section of the medium which at any time contains the same mass particles. Mathematically, $R(t)$ is defined as $R(t) = \{\mathbf{x}(t, \mathbf{a}) \, ; \, \mathbf{a} \in R(0)\}$, and a material region typically changes its shape and position as time passes, coinciding with the control volume at one instant of time, say at $t = 0$.

In this section we shall briefly review how the most important conservation laws of fluid mechanics may be derived by applying a simple result from vector analysis, namely *Reynolds' Transport Theorem*.

9.1 Reynolds Transport Theorem

From vector calculus we may know famous results such as the *Divergence Theorem* (also called Gauss' Theorem), and *Stokes' Theorem*. Reynolds Transport Theorem is another result in the same family. We introduce the theorem by first considering one-dimensional integrals.

DOI: 10.1201/9781003725206-9

If we need to take the derivative with respect to t of $H(t)$ defined by

$$H(t) = \int_{a(t)}^{b(t)} f(x,t)\, \mathrm{d}x,$$

it is possible first to write

$$H(t) = F(b(t),t) - F(a(t),t),$$

where F is the anti-derivative of f with respect to the first argument, and then apply the Chain Rule,

$$\frac{\mathrm{d}H}{\mathrm{d}t}(0) = \frac{\mathrm{d}}{\mathrm{d}t}\int_{a(0)}^{b(0)} f(x,t)\, \mathrm{d}x \bigg|_{t=0} + f(b(0),0)\frac{\partial b}{\partial t}(0) - f(a(0),0)\frac{\partial a}{\partial t}(0).$$

This result is useful to know, and we see that in addition to the expected first term, we have extra contributions because the integration interval changes with time.

Reynolds Transport Theorem is this identity when we are integrating over a moving region in space. We shall assume that the region we are looking at, $R(t)$, is enclosed by a moving boundary, $\partial R(t)$. Furthermore, we assume that the points on the boundary are marked so that we can trace them as time passes. In particular, all points on the boundary will, at any time, have a velocity $\mathbf{v}(\mathbf{x}(t))$, where $\mathbf{x}(t) \in \partial R(t)$.

We may then formulate Reynolds Transport Theorem for the integral of a function $\varphi(\mathbf{x},t)$ over the moving region $R(t)$ as follows:

$$\left(\frac{\mathrm{d}}{\mathrm{d}t}\int_{R(t)}\varphi(\mathbf{x},t)\mathrm{d}V\right)_{t=0} = \left(\frac{\mathrm{d}}{\mathrm{d}t}\int_{R(0)}\varphi(\mathbf{x},t)\mathrm{d}V\right)_{t=0} + \int_{\partial R(0)}\varphi(\mathbf{x},0)\mathbf{v}\cdot\mathbf{n}\mathrm{d}\sigma.$$

The theorem requires that $\mathbf{v}$ and φ are sufficiently nice functions, and that $R(t)$ is a nice region, but we will not go into that here. The proof follows directly from the definition of the derivative. We assume $R(0)$ and $R(t)$ are as outlined in Figure 9.1. Let $\Phi_i(t)$ denote the integral of φ over region "i"($i \in \{I, II, III\}$) at time t, that is,

$$\Phi_{III}(t) = \int_{III}\varphi(\mathbf{x},t)\mathrm{d}V.$$

In general, the regions **I** in **II** are defined by the parts of $\partial R(0)$ where the velocity field points in and out of $R(0)$, respectively. From the definition of the derivative and Figure 9.1 we have

$$\left(\frac{\mathrm{d}}{\mathrm{d}t}\int_{R(t)}\varphi(\mathbf{x},t)\mathrm{d}V\right)_{t=0} = \lim_{t\to 0}\frac{\Phi_{III}(t) + \Phi_{II}(t) - (\Phi_I(0) + \Phi_{III}(0))}{t}$$

$$= \lim_{t\to 0}\frac{\Phi_{I\cup III}(t) - \Phi_{I\cup III}(0)}{t} + \lim_{t\to 0}\frac{\Phi_{II}(t)}{t} - \lim_{t\to 0}\frac{\Phi_I(t)}{t}.$$

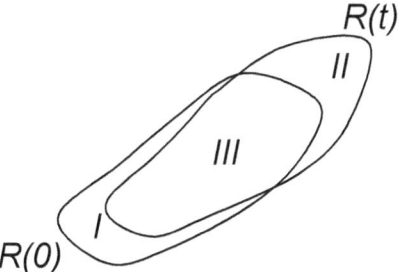

FIGURE 9.1
Definition of the regions I, II, and III.

The first limit value is just

$$\left(\frac{\mathrm{d}}{\mathrm{d}t}\int_{R(0)}\varphi(\mathbf{x},t)\mathrm{d}V\right)_{t=0}.$$

For small t, the parts **I** and **II** become thin shells such that we may use the volume elements $\mathrm{d}V = \mathbf{v}\cdot\mathbf{n}t\mathrm{d}\sigma$ for region **II**, and $\mathrm{d}V = -\mathbf{v}\cdot\mathbf{n}t\mathrm{d}\sigma$ for region **I**. In the limit $t\to 0$ we obtain

$$\lim_{t\to 0}\frac{\Phi_{\mathbf{II}}(t)-\Phi_{\mathbf{I}}(t)}{t} = \int_{\partial R(0)}\varphi(\mathbf{x},t)\mathbf{v}\cdot\mathbf{n}\mathrm{d}\sigma\bigg|_{t=0}.$$

In many textbooks, one finds that the theorem is stated assuming

$$\frac{\mathrm{d}}{\mathrm{d}t}\int_{R(0)}\varphi(\mathbf{x},t)\mathrm{d}V = \int_{R(0)}\frac{\partial\varphi}{\partial t}(\mathbf{x},t)\mathrm{d}V.$$

This is quite unfortunate for our applications and a direct error if φ has discontinuities inside $R(0)$ (recall the discussion in Sec. 7.3.4).

9.2 Mass Conservation

Mass conservation is a key principle in continuum mechanics. Here $\varphi = \rho$, the *mass* density of the medium. Without sources and sinks, the mass within a material region $R(t)$ will be constant, since this is precisely the definition of a material region. Thus,

$$\frac{\mathrm{d}}{\mathrm{d}t}\int_{R(t)}\rho(\mathbf{x},t)\mathrm{d}V = 0,$$

and, consequently, we get by applying the transport theorem

$$\frac{\mathrm{d}}{\mathrm{d}t}\int_{R(0)} \rho(\mathbf{x},t)\mathrm{d}V \bigg|_{t=0} + \int_{\partial R(0)} \rho(\mathbf{x},0)\mathbf{v}(\mathbf{x},0)\cdot\mathbf{n}(\mathbf{x},0)\mathrm{d}\sigma = 0.$$

Since there is nothing particular with time $t = 0$, we can for any time and an arbitrary fixed control volume R write

$$\frac{\mathrm{d}}{\mathrm{d}t}\int_R \rho\mathrm{d}V + \int_{\partial R} \rho\mathbf{v}\cdot\mathbf{n}\mathrm{d}\sigma = 0.$$

This is the mass conservation law in *integral form* when we do not have sources or sinks within R. Any sources/sinks will enter the expression as

$$\frac{\mathrm{d}}{\mathrm{d}t}\int_R \rho\mathrm{d}V + \int_{\partial R} \rho\mathbf{v}\cdot\mathbf{n}\mathrm{d}\sigma = \int_R q\mathrm{d}V.$$

Of course, we could see this immediately from the theory in Sec. 7.2, since mass is a material variable and the flux is $\rho\mathbf{v}$.

As discussed in Sec. 7.2, when ρ and v are sufficiently smooth, we can differentiate under the integral sign and apply the Divergence Theorem:

$$\frac{\mathrm{d}}{\mathrm{d}t}\int_R \rho\mathrm{d}V = \int_R \frac{\partial\rho}{\partial t}\mathrm{d}V,$$

$$\int_{\partial R} \rho\mathbf{v}\cdot\mathbf{n}\mathrm{d}\sigma = \int_R \nabla\cdot(\rho\mathbf{v})\mathrm{d}V,$$

so that

$$\int_R \left(\frac{\partial\rho}{\partial t} + \nabla\cdot(\rho\mathbf{v})\right)\mathrm{d}V = 0.$$

Holding for all R, this then leads to the differential formulation

$$\rho_t + \nabla\cdot(\rho\mathbf{v}) = 0.$$

The flow is called *stationary* if ρ and $\mathbf{v}$ are independent of time. Then $\int_R \rho\mathrm{d}V$ will be constant and the mass conservation reduces to

$$\int_{\partial R} \rho\mathbf{v}\cdot\mathbf{n}\mathrm{d}\sigma = 0. \tag{9.1}$$

Equation (9.1) can be directly used for calculations, as illustrated for the pipeline in Figure 9.2. By letting R be as given in the figure, the conservation

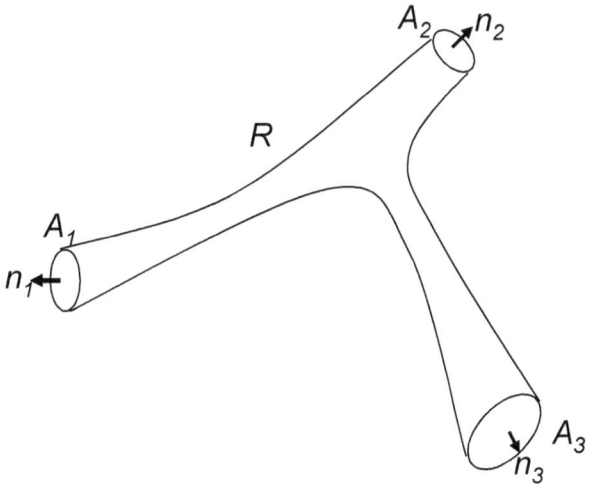

FIGURE 9.2
Stationary flow in a network of pipes will obey the mass conservation.

law when the flow is stationary is

$$\int_{A_1} \rho \mathbf{v} \cdot \mathbf{n} d\sigma + \int_{A_2} \rho \mathbf{v} \cdot \mathbf{n} d\sigma + \int_{A_3} \rho \mathbf{v} \cdot \mathbf{n} d\sigma = 0.$$

By assuming that ρ is constant and defining the velocity over the cross section of the tubes,

$$v_i = \frac{1}{|A_i|} \int_{A_i} \mathbf{v} \cdot \mathbf{n} d\sigma,$$

we obtain

$$|A_1| \rho_1 v_1 + |A_2| \rho_2 v_2 + |A_3| \rho_3 v_3 = 0.$$

($|A_j|$ is the cross-sectional area). As one understands, it is easy to generalize this to arbitrary networks and otherwise include sources and sinks.

We immediately deduce the following special cases of the differential formulation:

- Stationary flow (no time variation): $\nabla \cdot (\rho \mathbf{v}) = 0$;

- Constant density (*incompressible flow*): $\nabla \cdot \mathbf{v} = 0$.

9.3   Momentum Conservation

In mechanics, *momentum* (or *linear momentum*) is the product of a body's
mass times its velocity. Besides mass conservation, the momentum conservation law is the most important.

For a continuum, the *momentum density* is defined as momentum per unit
volume, $\mathbf{p} = \rho\mathbf{v}$, where ρ is the mass density and $\mathbf{v}$ the velocity. The density
is thus a three-dimensional vector in space. Newton's Second Law, stating
that force is equal to mass times acceleration, is a statement about a fixed
collection of mass particles. Since a material region $R(t)$ always contains the
same mass particles, Newton's law applied to a material region $R(t)$ is just

$$\frac{\mathrm{d}}{\mathrm{d}t} \int_{R(t)} \rho\mathbf{v}\mathrm{d}V = \Sigma\mathbf{F}(t),$$

where $\Sigma\mathbf{F}(t)$ is the sum of all forces acting on the material region. By applying
the Reynolds transport theorem at $t = 0$ we obtain

$$\left(\frac{\mathrm{d}}{\mathrm{d}t} \int_{R(0)} \rho\mathbf{v}\mathrm{d}V + \int_{\partial R(0)} (\rho\mathbf{v})\,(\mathbf{v}\cdot\mathbf{n})\,\mathrm{d}\sigma \right)_{t=0} = \Sigma\mathbf{F}(0).$$

For a fixed control volume R, the conservation of momentum may therefore
be expressed as

$$\frac{\mathrm{d}}{\mathrm{d}t} \int_{R} \rho\mathbf{v}\mathrm{d}V + \int_{\partial R} (\rho\mathbf{v})\,(\mathbf{v}\cdot\mathbf{n})\,\mathrm{d}\sigma = \Sigma\mathbf{F}. \tag{9.2}$$

The equation can be used to express the conservation of momentum in any
direction: If $\mathbf{a}$ is a fixed unit vector in space, the scalar product with (9.2)
gives

$$\frac{\mathrm{d}}{\mathrm{d}t} \int_{R} \rho v_a\mathrm{d}V + \int_{\partial R} (\rho v_a)\,(\mathbf{v}\cdot\mathbf{n})\,\mathrm{d}\sigma = \Sigma F_a,$$

where $v_a = \mathbf{a}\cdot\mathbf{v}$ and $F_a = \mathbf{a}\cdot\mathbf{F}$. For a Cartesian coordinate system in space,
the three standard unit vectors give us three equations corresponding to the
axes.

To get further, it is necessary to say something about the forces acting on
the mass in R. It is common to distinguish between *mass forces* (also called
body forces) and *surface forces*. In general, it is possible to write

$$\mathbf{F}_B = \int_{R} \mathbf{f}_B(\mathbf{x}, t)\mathrm{d}V,$$

for mass forces and

$$\mathbf{F}_S = \int_{\partial R} \mathbf{f}_S(\sigma, t) \mathrm{d}\sigma,$$

for surface forces.

The most common mass force is gravity,

$$\mathbf{F}_g = \int_R \rho \mathbf{g} \mathrm{d}V.$$

In the geophysical fluid flow (such as oceans and atmosphere), the *Coriolis force* and the *centripetal acceleration* are important. These are forces that arise because our control volume is fixed on earth's surface and is thus rotating with the earth. The Coriolis force is given by

$$\mathbf{F}_c = \int_R \rho(-2\mathbf{\Omega} \times \mathbf{v}) \mathrm{d}V,$$

where $\mathbf{\Omega}$ is the angular velocity of the earth ($2\pi/24$ hours $= 7.3 \times 10^{-5} \mathrm{s}^{-1}$). The centripetal acceleration

$$\mathbf{F}_s = \int_R \rho(-\mathbf{\Omega} \times (\mathbf{\Omega} \times \mathbf{r})) \mathrm{d}V,$$

where $\mathbf{r}$ is the position vector from the centre of the earth. Electromagnetic forces are other important examples of body forces.

The forces acting on the surface of R may be expressed in terms of the so-called *stress tensor* of the medium. Stress is the force per unit area. The force can act along a surface (*shear stress*) or orthogonal to the surface (*normal stress*). We refer to courses in mechanics for further discussion of the stress tensor. If we equip space with a Cartesian coordinate system, we may represent the stress tensor by a symmetric 3×3 matrix,

$$\mathbf{T} = \begin{bmatrix} t_{11} & t_{12} & t_{13} \\ t_{21} & t_{22} & t_{23} \\ t_{31} & t_{32} & t_{33} \end{bmatrix}.$$

For a small surface element $\mathrm{d}\sigma$ with normal vector $\mathbf{n}$, the force acting on $\mathrm{d}\sigma$ is given by

$$\mathrm{d}\mathbf{F} = \mathbf{T} \cdot \mathbf{n} \mathrm{d}\sigma,$$

and the momentum conservation law in integral form may be written

$$\frac{\mathrm{d}}{\mathrm{d}t} \int_R \rho \mathbf{v} \mathrm{d}V + \int_{\partial R} (\rho \mathbf{v})\mathbf{v} \cdot \mathbf{n} \mathrm{d}\sigma = \int_R \rho \mathbf{f}_B \mathrm{d}V + \int_{\partial R} \mathbf{T} \cdot \mathbf{n} \mathrm{d}\sigma.$$

If we are working with liquids, the stress tensor has contributions from pressure and viscosity forces. The pressure acts orthogonal on a small area element within the fluid and has the same value at a point no matter how the element is oriented. In addition, all common liquids are more or less viscous. Viscosity can be seen as a type of internal friction which provides resistance against deformations. For so-called *Newtonian fluids*, the shear stress in the x-direction for a flow with velocity $u(y)$ parallel to the x-axis is given by

$$\tau = \mu \frac{\partial u}{\partial y}, \tag{9.3}$$

where μ is called *dynamic viscosity*. It may be shown (See, *e.g.* [28]) from the mathematical properties of the stress tensor that the simplest expression consistent with the equation (9.3) and giving the static pressure p when the fluid is at rest, has to be of the form

$$t_{ij} = \left(-p - \frac{2}{3}\mu \nabla \cdot \mathbf{v} \right) \delta_{ij} + \mu \left(\frac{\partial v_i}{\partial x_j} + \frac{\partial v_j}{\partial x_i} \right), \quad i,j = 1,2,3,$$

(indices refer to the standard Cartesian coordinate system). This expression leads to the quite famous equation for the momentum balance in a Newtonian fluid,

$$\frac{\partial}{\partial t}\rho \mathbf{v} + \nabla \cdot ((\rho\mathbf{v})\mathbf{v}) = \rho\mathbf{f}_B - \nabla p + \mu \left(\nabla^2 \mathbf{v} + \frac{1}{3}\nabla(\nabla \cdot \mathbf{v}) \right),$$

called *Navier-Stokes Equation(s)*. If the liquid is incompressible and the density is constant, the equation simplifies to

$$\frac{\partial}{\partial t}\mathbf{v} + \nabla \cdot ((\mathbf{v})\mathbf{v}) = \mathbf{f}_B - \frac{1}{\rho}\nabla p + \frac{\mu}{\rho}\nabla^2 \mathbf{v},$$

since $\nabla \cdot \mathbf{v} = 0$.

9.4 Energy Conservation

The first law of thermodynamics says that for a system in thermodynamic equilibrium, the added heat will be used to perform work and change the system's internal energy, that is,

$$dQ = dW + dE.$$

The energy may be expressed as specific energy e (energy per unit mass) so that

$$E(t) = \int_{R(t)} e\rho dV.$$

Contrary to Q and W, the specific energy e is a material variable. Specific energy may consist, for example, of *kinetic* and *inner* energy per mass unit,

$$e = \mathbf{v} \cdot \mathbf{v}/2 + u.$$

Work performed by the system may be of different kinds. If we consider the work per unit time (power), we have

(i) *Work against the mass forces:*

$$\frac{\mathrm{d}W_B}{\mathrm{d}t} = - \int_R \mathbf{f}_B \cdot \mathbf{v} \mathrm{d}V$$

(ii) *Work against surface forces:*

$$\frac{\mathrm{d}W_S}{\mathrm{d}t} = - \int_{\partial R} (\mathbf{T} \cdot \mathbf{n}) \cdot \mathbf{v} \mathrm{d}\sigma$$

(iii) *Other work performed by the system (e.g., driving a turbine):*

$$\frac{\mathrm{d}W_t}{\mathrm{d}t}.$$

From the first law and Reynolds transport theorem, the general energy conservation law becomes

$$\frac{\mathrm{d}}{\mathrm{d}t} \int_R e\rho \mathrm{d}V + \int_{\partial R} e\rho \mathbf{v} \cdot \mathbf{n} \mathrm{d}\sigma = \frac{\mathrm{d}Q}{\mathrm{d}t} - \frac{\mathrm{d}W_t}{\mathrm{d}t} + \int_R \mathbf{f}_B \cdot \mathbf{v} \mathrm{d}V + \int_{\partial R} (\mathbf{T} \cdot \mathbf{n}) \cdot \mathbf{v} \mathrm{d}\sigma.$$

In the same way as for mass conservation, we can also derive the differential formulation using the Divergence Theorem, provided that the smoothness conditions are fulfilled. Further information may be found in textbooks about continuum mechanics.

9.5   Comments and Examples

There are several other conservation laws besides those presented here. In particular, the conservation law for *vorticity* ($\nabla \times \mathbf{v}$) is important for many applications in fluid mechanics.

Traditional mechanics and mathematics teaching are oriented towards differential equations, i.e., differential formulations. This is natural since there is a huge theory about the existence of solutions, and techniques such as separation of variables, integral transforms, Green functions, and perturbation methods for finding solutions.

Nevertheless, modern textbooks of practical mechanics to a greater extent base their arguments on integral formulations. The integral formulations are independent of the choice of coordinate system and embody the fundamental physical laws (which, after all, manage the real world) more directly than differential formulations. Integral formulations may be used for practical tasks and a control volume need not be just a small box!

Conservation laws also apply to situations where differential equations have shortcomings, such as for discontinuous variables. This is especially important for treating shock solutions.

While numerical models traditionally have been made from differential equations by replacing the derivatives with the finite difference approxima-tions, one can also use the integral formulation directly by dividing the com-putational region into a pile of boxes. The equations for each box are then established based on the conservation laws. This guarantees that the numerical solutions are compatible with the conservation laws. Finite Element formula-tions and so-called weak solutions of differential equations are also related to the conservation laws in integral form.

Below we will look at three examples of how one can operate with con-servation laws. The first example should be familiar to everyone with some background in fluid mechanics.

The second example deals with the phenomenon of shock, and is typical for that type of problems. The conservation laws provide conditions that help us to determine the properties of the shock. In aerodynamics shocks are associated with supersonic speeds, while the hydrodynamic shock in the example occurs at the very mundane speeds. This is also the case for the third, somewhat more challenging, example.

9.5.1   Forces on a pipe bend

We consider a tube bend with stationary horizontal flow (see Figure 9.3). We know the pressure, the cross-sectional area, density, and the speed at both the inlet and outlet. The problem is to find the forces F_x and F_y that we must apply in order to keep the bend in position. The velocities are vectors with directions indicated by arrows, and we assume that the velocity magnitude and the pressure are constant over the cross sections (A_1 and A_2). Since the flow is stationary, the mass conservation requires

$$\rho_1 v_1 A_1 = \rho_2 v_2 A_2.$$

For the momentum balance, we must first get an overview of the forces on R, which, in addition to the force needed to keep the bend in position, are composed of pressure forces:

$$(\Sigma F)_x = p_1 A_1 - F_x - p_2 A_2 \cos \alpha,$$
$$(\Sigma F)_y = -F_y + p_2 A_2 \sin \alpha.$$

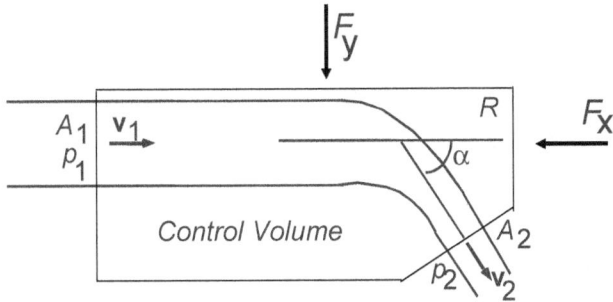

FIGURE 9.3
How much force should we apply to keep a bent tube in position?

Since conditions are stationary,

$$\frac{\mathrm{d}}{\mathrm{d}t} \int_R \rho \mathbf{v} \, \mathrm{d}V = 0,$$

whereas

$$\int_{\partial R} \rho v_x \mathbf{v} \cdot \mathbf{n} \, \mathrm{d}\sigma = \rho_1 v_1 (-v_1 A_1) + \rho_2 v_2 \cos \alpha (v_2 A_2),$$

$$\int_{\partial R} \rho v_y \mathbf{v} \cdot \mathbf{n} \, \mathrm{d}\sigma = \rho_1 \cdot 0 \cdot (-v_1 A_1) + \rho_2 (-v_2 \sin \alpha)(v_2 A_2).$$

If this is inserted into the conservation law, we obtain

$$F_x = p_1 A_1 - p_2 A_2 \cos \alpha + \rho_1 v_1^2 A_1 - \rho_2 v_2^2 A_2 \cos \alpha,$$
$$F_y = p_2 A_2 \sin \alpha + \rho_2 v_2^2 A_2 \sin \alpha.$$

Since $\rho_1 v_1 A_1 = \rho_2 v_2 A_2 = M$, we get

$$F_x = M(v_1 - v_2 \cos a) + p_1 A_1 - p_2 A_2 \cos \alpha,$$
$$F_y = M v_2 \sin \alpha + p_2 A_2 \sin \alpha.$$

9.5.2 Flood waves in rivers

In this example we shall look at the simplest theory of flood waves and water jumps in rivers. Since the water level behind the water jump is higher than the level in front of the jump, people and livestock along the river can be swept away by the water, or suddenly find themselves in much deeper water than they appreciate. The reason for the jump could be torrential rain, sudden emission of water from a power station or a dam break.

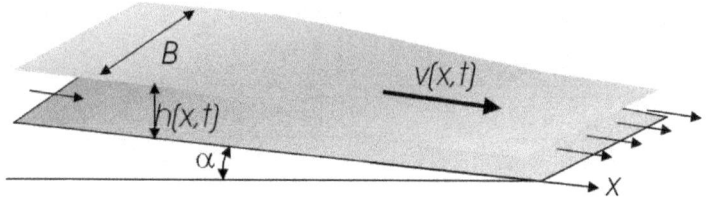

FIGURE 9.4
An idealized picture of a river.

To model what happens, we shall consider water flowing down a slope with a relatively small tilt angle α. The width of the flow across the inclined plane is B.

We assume that the velocity of the water is directed down the slope and has value v, is a function of position x and time t, and independent of z and y. This assumes that the flow is turbulent and that the water depth is not too large. In that case, the velocity is approximately constant over depth, but close to the bottom we will have a *boundary layer* where this assumption is not so good. Furthermore, we assume that the surface is defined by $z = h(x,t)$. To model the flow, we use mass and momentum conservation. The situation is illustrated in Figure 9.4. Let us first see what mass conservation provides. The conservation law has the general form

$$\frac{d}{dt}\int_R \rho dV + \int_{\partial R} \mathbf{j} \cdot \mathbf{n} d\sigma = 0. \tag{9.4}$$

We assume that the density ρ is constant and use $dV = Bh(x,t)dx$. The amount j that flows past a point x per unit time has the form

$$j(x,t) = \rho v(x,t) \cdot (Bh(x,t)).$$

The flux vector (amount per surface and unit time) is thus as expected

$$\mathbf{j}(x,t) = \rho v(x,t)\,\hat{\imath}.$$

This may be inserted into (9.4) for a section of the river between $x = a$ and $x = b$, while using $\mathbf{n} = -\hat{\imath}$ at $x = a$ and $\mathbf{n} = \hat{\imath}$ at $x = b$:

$$0 = \frac{d}{dt}\int_a^b \rho Bh(x,t)dx + [(\rho v)(Bh)]_a^b$$

$$= \frac{d}{dt}\int_a^b \rho Bh(x,t)dx + [\rho v(b,t) \cdot Bh(b,t) - \rho v(a,t) \cdot Bh(a,t)] = 0.$$

Note that the flux orthogonal to the bottom and the surface is zero, and also in the y-direction, since we assume no flow in that direction. By letting $a \to b$, dividing by $b-a$ in the usual way, and moving the derivative inside the integral sign, we obtain the equation in differential form:

$$\frac{\partial h}{\partial t} + \frac{\partial}{\partial x}(vh) = 0. \tag{9.5}$$

Since this is an equation with two unknown functions, v and h, we can not solve the equation immediately.

The momentum balance will here give us something only for the x-direction, $p_x = \rho v$, and the conservation law takes the following form:

$$\frac{\mathrm{d}}{\mathrm{d}t} \int_a^b (\rho v) \, Bh \mathrm{d}x + [(\rho v) \, v \cdot (Bh)]_a^b = \sum F_x.$$

It remains to specify the forces. Gravity acts directly on the water and, along the x-axis, this amounts to the force component proportional to $\sin \alpha$, that is,

$$F_g = \int_a^b \rho g \sin \alpha B h \, (x,t) \, \mathrm{d}x.$$

We then have the pressure forces. The pressure at the surface and the bottom does not contribute significantly to the x-component. However, we have a contribution from the end surfaces. Here we shall assume hydrostatic pressure, $p = \rho g \, (h - z)$, and set $\cos \alpha \approx 1$ so that

$$\mathrm{d}P = \rho g \, (h - z) \, (B \mathrm{d}z).$$

The total pressure force on the surface at $x = a$ is thus

$$P \, (a,t) = \int_0^h \rho g \, (h - z) \, B \mathrm{d}z = \rho g B \frac{h^2 \, (a,t)}{2},$$

and similarly at $x = b$, where the pressure force acts in the negative x-direction,

$$P \, (b,t) = -\rho g B \frac{h^2 \, (b,t)}{2}.$$

The final force contribution is the friction force against the bottom. It turns out, partly based on dimensional analysis, one can assume the friction force per unit area (*shear stress* in the x-direction) to be

$$\tau = -\rho C_f \mathbf{v} \, |\mathbf{v}|.$$

The constant C_f is called the *Chézy factor* and is empirically determined and depending on the roughness of the bottom. The total friction force is therefore found by integrating τ over the bottom surface:

$$F_f = -\int_a^b \rho C_f v^2 \, (B \mathrm{d}x).$$

If we put all this together and divide by ρB, we get

$$\frac{\mathrm{d}}{\mathrm{d}t} \int_a^b vh\,\mathrm{d}x + \left[(v)\,vh + \frac{g}{2}h^2\right]_a^b = \int_a^b \left(g\sin\alpha h - C_f v^2\right)\,\mathrm{d}x.$$

It may be a bit tedious to establish the conservation law, but the principle is simple. The differential formulation follows in the same way as above:

$$\frac{\partial(hv)}{\partial t} + \frac{\partial}{\partial x}\left(v^2 h + \frac{g}{2}h^2\right) = g\sin\alpha h - C_f v^2. \tag{9.6}$$

The equations (9.5) and (9.6) are often called the *shallow water equations*, or *Saint-Venant Equations*, and constitute what is called a hyperbolic system. There is a theory for hyperbolic systems of two equations that we shall not go into here, but in general, the equations cannot be solved analytically. However, it is easy to see that the equations have the solution

$$h(x,t) = h_0,$$
$$v(x,t) = v_0,$$

where

$$g\sin\alpha h_0 - C_f v_0^2 = 0.$$

The last equation simply says that friction balances gravity. Linear stability analysis can tell whether the solution is stable, and this analysis, which is analogous to the one made for the instabilities in a traffic jam, may be found in the book by Whitham [41], pp. 85–86.

In certain situations, the flow is unstable, and more advanced analysis leads to so-called *roll waves*. Roll waves may be observed on smooth sloping surfaces during heavy rain. Water flowing down a flat slope then has a tendency to create "waves" that are almost vertical in front and move slowly downward in relation to water velocity itself.

If we neglect the left side in (9.6), gravity always balances the friction. This is called the *kinematic theory* of flood waves. We obtain a relation between h and v,

$$v = \sqrt{\frac{g\sin\alpha}{C_f}}\,h^{1/2},$$

and (9.5) becomes

$$\frac{\partial h}{\partial t} + \sqrt{\frac{g\sin\alpha}{C_f}}\frac{\partial}{\partial x}h^{3/2} = 0.$$

The kinematic velocity is

$$c(h) = \frac{\mathrm{d}}{\mathrm{d}h}\sqrt{\frac{g\sin\alpha}{C_f}}\,h^{3/2} = \frac{3}{2}\sqrt{\frac{g\sin\alpha}{C_f}}\,h^{1/2} = \frac{3}{2}v(h).$$

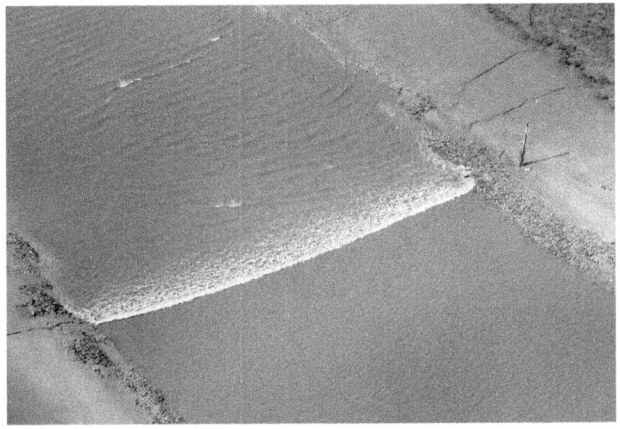

FIGURE 9.5
Tidal bores see from the River Ribble Lancashire (Image copied from `https://en.m.wikipedia.org/wiki/File:River_Ribble_bore.jpg` ; licensed under the Creative Commons Attribution-Share Alike 3.0 Unported license).

We leave it to the reader to show that if the water level in the upper part of a river increases, in other words, if $\partial h/\partial x < 0$, it may develop a shock that in this case moves down the river like a wall. The phenomenon can occur during torrential rain or in rivers with regulated water flow, such as in rivers downstream from power plants. Note that the speed of the shock will be about 50% greater than the speed of the water flow!

On the Figure 9.5, we see an example of similar phenomena. These waves are called *tidal bores*, and occur as shock-solutions for the equations (9.5) and (9.6) in a flat river when the tide enters the river from its mouth.

9.5.3 Research project: The circular water jump

Everyone who has run tapped water vertically into the kitchen sink has observed that a circular water jump often forms some distance from where the jet hits the surface. If this does not sound familiar, one should before reading further make a simple experiment as in Figure 9.6. The geometry of the problem is indicated in Figure 9.7. We assume radial symmetry and a constant density ρ. Furthermore, we assume that the water velocity is directed radially outward and is independent of z. Thus, both speed and depth are only functions of r and t. As for the pressure, we assume that this is given by the hydrostatic pressure, $p(r, t, z) = \rho g(h(r, t) - z)$ since any constant atmospheric pressure drops out. Frictional force per area unit at the bottom has the form $\mathbf{t}_C = -C_f \rho |\mathbf{v}| \mathbf{v}$ (where again C_f is the Chezy friction factor).

Let us set up the general conservation laws of mass and momentum for a control volume limited by r_1 and r_2, $r_1 < r_2$, where r_1 is greater than the

FIGURE 9.6
Circular hydraulic jump in a sink formed some distance away from where the stream hits. (Image copied from `https://en.m.wikipedia.org/wiki/File:Hydraulic_jump_in_sink.jpg`).

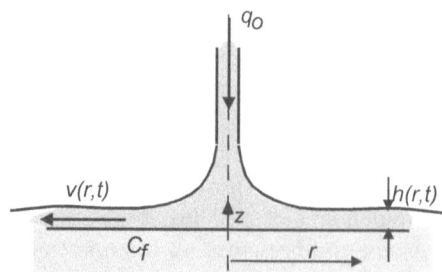

FIGURE 9.7
Vertical water jet hitting a horizontal plane.

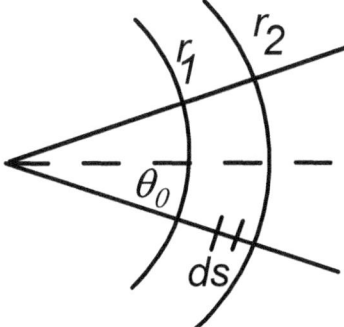

FIGURE 9.8
Sector-shaped control volume for the momentum balance.

radius of the centre jet. From the mass conservation law,

$$\frac{d}{dt} \int_R \rho dV + \int_{\partial R} \rho (\mathbf{v} \cdot \mathbf{n}) \, d\sigma = 0.$$

We immediately get, using polar coordinates and dividing by $2\pi\rho$,

$$\frac{d}{dt} \int_{r=r_1}^{r_2} h(r,t)rdr + r_2 h_2 v_2 - r_1 h_1 v_1 = 0.$$

For the momentum balance, it is necessary to choose a pie-shaped section as control volume and compute, *e.g.*, the momentum balance in the x-direction (see Figure 9.8). Since the pressure on the bottom surface works vertically; it is sufficient to look at the pressure forces on the side walls. The pressure forces acting on a strip of width ds of the side wall is $dP = \rho g h^2(r,t)ds/2$, and by integrating around all side walls we obtain

$$P_x = \int_{\text{Sides}} -pn_x d\sigma = \rho g \sin\theta_0 \left(h^2(r_1,t)r_1 - h^2(r_2,t) + \int_{r=r_1}^{r_2} h^2(r,t)dr \right).$$

For the bottom friction we have, as for the flood waves,

$$C_x = - \int_{r=r_1}^{r_2} \int_{\theta=-\theta_0}^{\theta_0} C_f \rho v \cdot (v \cos\theta) rdrd\theta = -2C_f \rho \sin\theta_0 \int_{r=r_1}^{r_2} v^2(r,t)rdr.v.$$

(9.7)

The rest of the expressions is left to the reader, and we finally end with

$$\frac{d}{dt} \int_{r_1}^{r_2} vhrdr + \left[v^2rh + \frac{r}{2}h^2 g \right]_{r_1}^{r_2} = \int_{r_1}^{r_2} \left(-C_f v^2 r + \frac{h^2}{2}g \right) dr.$$

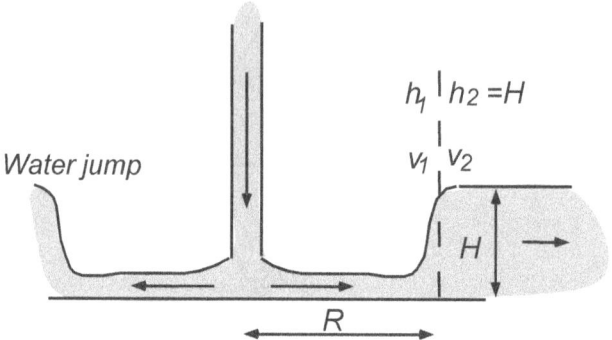

FIGURE 9.9
Formation of a circular water jump.

Differential formulations of mass and momentum follow in the usual way:

$$\frac{\partial(rh)}{\partial t} + \frac{\partial(rhv)}{\partial r} = 0,$$

$$\frac{\partial(rhv)}{\partial t} + \frac{\partial}{\partial r}(rhv^2 + rh^2g/2) = -C_f v^2 r + \frac{h^2}{2}g.$$

A suitable outflow will produce a stationary annular jump some distance from where the jet strikes, as indicated in Figure 9.9. The water jump is actually a shock called a *hydraulic jump*. By assuming steady flow and letting $r_1 \to r_2 = R$ in the conservation laws, all the integrals of the conservation laws disappear, and we end up with the following classical shock conditions derived by J.-B. Belanger in 1838 [39]:

$$v_1 h_1 = v_2 h_2. \tag{9.8}$$

$$v_1^2 h_1 + \frac{1}{2}h_1^2 g = v_2^2 h_2 + \frac{1}{2}h_2^2 g. \tag{9.9}$$

This is not sufficient to determine the position of the shock. We refer to the fluid mechanics textbooks for a discussion of the energy conservation in an ideal fluid under stationary conditions, leading to *Bernoulli's equation*, stating that the quantity

$$gz + \frac{v^2}{2} + \frac{p}{\rho}$$

is constant along streamlines. For a streamline on the surface (or at the bottom) this gives a third condition at the jump:

$$gh_1 + \frac{v_1^2}{2} = gh_2 + \frac{v_2^2}{2}. \tag{9.10}$$

In the theory of hyperbolic conservation laws, conditions such as the ones stated in the equations (9.8), (9.9), and (9.10) are called *Rankine-Hugoniot conditions*. These equations have the obvious and not particularly exciting solution $h_1 = h_2$, $v_1 = v_2$, and $h_L = 0$. However, if one requires that $h_1 \neq h_2$, the three equations do not have a solution. In reality, some of the energy is transformed to turbulence and eventually to heat at the shock. This requires the energy condition to include a certain energy loss at the jump:

$$gh_1 + \frac{v_1^2}{2} = gh_2 + \frac{v_2^2}{2} + gh_L.$$

The extra quantity gh_L on the right side is called *head loss* and is basically unknown.

If we only consider (9.8) and (9.9), it is possible to derive the equation

$$\left(\frac{h_2}{h_1}\right)^2 + \frac{h_2}{h_1} = 2\frac{v_1^2}{gh_1}.$$

The dimensionless combination

$$Fr = \frac{v}{\sqrt{gh}}$$

occurring on the right side is called *Froude Number*, and is an important number in hydrodynamics. The quadratic equation for h_2/h_1 has the solution

$$\frac{h_2}{h_1} = \frac{1}{2}\left(-1 \pm \sqrt{1 + 8Fr_1^2}\right) \qquad (9.11)$$

For h_1 to be less than h_2, it is necessary that $Fr_1 > 1$. The Froude number has an interesting physical interpretation. Water waves (longer than a few centimeters) have the propagation speed $c_p = \sqrt{gh}$ in shallow water. Thus, the Froude number is the ratio between v and c_p. A flow with a free surface where $v > c_p$ is called *supercritical flow*. If you are sitting in the flow, a disturbance in front of you cannot be warned by a surface wave before it happens to you. Thus, Fr is analogous to the *Mach number* in aerodynamics.

From the energy condition, we can find an expression for the relative energy loss in the shock expressed as the ratio h_L/h_1:

$$\frac{h_L}{h_1} = \frac{h_1 - h_2 + v_1^2/2g - v_2^2/2g}{h_1},$$
$$= 1 - \frac{h_2}{h_1} + \frac{Fr^2}{2} - \frac{v_2^2}{2gh_1} = 1 - \frac{h_2}{h_1} + \frac{Fr^2}{2}\left(1 - \left(\frac{1}{h_2/h_1}\right)^2\right),$$

where we finally may insert (9.11).

But what is the position of the water jump? A simplified analysis can be found in [39], ignoring all other energy losses apart from that at the shock. If the kinetic energy in the flow before the shock is much larger than potential energy, the velocity, according to Bernoulli's equation, will be approximately constant and equal to U_0, i.e., the speed of the jet as it hits the plate. We then obtain from the mass and momentum balance,

$$q_0 = 2\pi R h_1 U_0 = 2\pi R H v_2,$$
$$U_0^2 h_1 + h_1^2 g/2 = v_2^2 H + H^2 g/2,$$

which, after a simple transformation, gives

$$R = \frac{(U_0^2 - gH/2)q_0}{\pi g H^2 U_0}.$$

It is not unreasonable that U_0 enters in addition to q_0, since both the added momentum and mass should be of importance for the position of the jump.

In September 1993, *Journal of Fluid Mechanics* presented a comprehensive analysis of the problem [9].

10

Diffusion and Convection

Whereas a material variable is a quantity passively transported along with the flowing medium, this is not always a reasonable assumption. On a small scale, *molecular diffusion* mixes liquids and gases by the molecules tumbling around. This also applies if we add small particles to the fluid (mixing by *Brownian motion*). On a larger scale, there are differences in concentrations that give rise to the mixing. If the fluid is moving, there will also be a change because of transportation. Such transport is called *convection*.

Low viscosity fluids will often have speeds and length scales so that *turbulence* is developed. Turbulence is the chaotic movement where the flow develops vortices and thin layers that spin into each other and split up (smoke in the air visualizes this well). If we add a foreign substance to a medium, we observe that the mixing goes much faster when the flow is turbulent than it does if the mixture occurs by diffusion alone (this is why we use a spoon in the cup to stir up milk and coffee!). Modelling of turbulent mixing is a classical modelling problem. The simplest would be trying to describe it as an enhanced molecular diffusion, but there are also examples where such a description appears to be quite wrong. Turbulence in fluids can occur when the speed varies greatly from place to place (called velocity shear in fluid mechanics). Otherwise, turbulence may be caused by temperature and density differences (always observed for boiling water). In Section 10.6, we shall derive the conservation laws for turbulence from the conservation laws in Chapter 9.

In addition to molecular diffusion and turbulence, diffusion-like spread has shown to be very applicable in other contexts. Often, the offspring is a discrete phenomenon, modelled by stochastic random walk. Mathematically, one can then show that such discrete models in the limit of very many objects transform into diffusion models.

10.1   Conservation Laws with Diffusion

Diffusion is flux caused by *concentration differences*. If the concentration is constant, there will be no net flux in any direction. To first order, the flux must be proportional to the change in concentration per unit length, in other

DOI: 10.1201/9781003725206-10

words

$$\mathbf{j} = -\sigma \nabla \varphi,$$

where φ is the concentration and σ is called the *diffusion coefficient*. This expression is called *Fick's Law* of diffusion. Heat conduction obeys a similar law. If the specific heat is constant, the heat flux is $\mathbf{q} = -k \nabla T$, where T is the temperature and k is called the *heat conduction coefficient*. This expression is called *Fourier's Heat Conduction Law*.

Diffusion has a *smoothing* effect on the concentration gradients. If we start with a localized section R of a liquid with a different concentration of some substance, the largest concentration gradients occur along the edges of R. When we stir and develop turbulence, the batch of liquid is stretched and twisted so that an increasing proportion is located in areas with strong gradients. In this way, diffusion acts more strongly, and this is the mechanism behind what we call *forced mixing* or enhanced diffusion by turbulence.

Let us now consider a general situation where $c(\mathbf{x}, t)$ is the concentration of a substance in a liquid, and the vector field $\mathbf{j}(\mathbf{x}, t)$ denotes the corresponding flux. If the substance passively follows the flow, we have shown before that the flux is just $\mathbf{j}_c(\mathbf{x}, t) = c\mathbf{v}$. Even when the liquid is at rest, the material may be spread by a diffusion flux $\mathbf{j}_d$. We want to show that the total flux will be the sum of the fluxes, $\mathbf{j}(\mathbf{x}, t) = \mathbf{j}_d(\mathbf{x}, t) + \mathbf{j}_c(\mathbf{x}, t)$. Let $R(t)$ be a material region of the liquid much smaller than the scale of variations in $\mathbf{v}$. For time intervals of the order of the diameter of R divided by $|\mathbf{v}|$, an observer travelling with $R(t)$ set up

$$\frac{\mathrm{d}}{\mathrm{d}t} \int_{R(t)} c(\mathbf{x}, t)\, \mathrm{d}V \bigg|_{t=0} + \int_{\partial R(0)} j_d(\mathbf{x}, 0) \cdot \mathbf{n}\mathrm{d}\sigma = \int_{R(0)} q(\mathbf{x}, t)\, \mathrm{d}V.$$

But Reynolds transport theorem applied to the first term gives

$$\frac{\mathrm{d}}{\mathrm{d}t} \int_{R(t)} c(\mathbf{x}, t)\, \mathrm{d}V \bigg|_{t=0} = \frac{\mathrm{d}}{\mathrm{d}t} \int_{R(0)} c(\mathbf{x}, t)\, \mathrm{d}V + \int_{\partial R(0)} c(x, 0)\, \mathbf{v} \cdot \mathbf{n}\mathrm{d}\sigma,$$

and altogether,

$$\frac{\mathrm{d}}{\mathrm{d}t} \int_R c\mathrm{d}V + \int_{\partial R} (\mathbf{j}_d + c\mathbf{v}) \cdot \mathbf{n}\mathrm{d}\sigma = \int_R q\mathrm{d}V.$$

A general control region can be divided into arbitrarily small parts where this formula holds. When all contributions are added, the area integrals over common borders cancel so that one ends up with the same formula for the whole region. This is thus the general diffusion conservation law, and the total flux is

$$\mathbf{j}(\mathbf{x}, t) = \mathbf{j}_d(\mathbf{x}, t) + \mathbf{j}_c(\mathbf{x}, t).$$

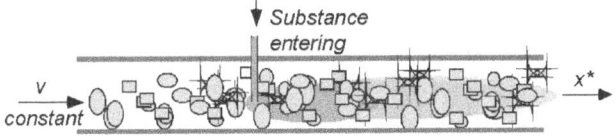

FIGURE 10.1
Sketch of a simple one-dimensional chemical reactor. The fluid flows through
a tube filled with crushed glass or glass spheres.

If we apply the Divergence Theorem as in Chapter 9, we find the differential
form:
$$\frac{\partial c}{\partial t} + \nabla \cdot (\mathbf{j}_d + \mathbf{v}c) = q.$$

Inserting for $\mathbf{j}_d = -\sigma \nabla c$, we get a *Convection/Diffusion Equation*:

$$\frac{\partial c}{\partial t} + \nabla \cdot (\mathbf{v}c) - \nabla \cdot (\sigma \nabla c) = q. \tag{10.1}$$

In general, both $\mathbf{v}$ and σ may depend on c so that the equation is nonlinear.
In the next section we will consider a simple mathematical model where this
equation enters in a central way.

10.2   One-Dimensional Chemical Reactor

A simple chemical reactor consists of a tube (often filled with crushed glass or
glass spheres) where a fluid is flowing with constant mean velocity v. A certain
substance is added to the fluid from a nozzle. The substance is mixed into the
fluid with a constant diffusion coefficient σ. In practice, σ is the *effective*
diffusion coefficient due to the turbulent mixing in the flow around the glass
obstacles), as illustrated in Figure 10.1. The concentration of the substance
(outside the source region) is described by a one-dimensional version of (10.1):

$$\frac{\partial c}{\partial t^*} + v\frac{\partial c}{\partial x^*} - \sigma \frac{\partial^2 c}{\partial x^{*2}} = 0.$$

Scaling x^* by the typical tube length L and a corresponding time scale $T = L/v$ gives us

$$\frac{\partial c}{\partial t} + \frac{\partial c}{\partial x} - \varepsilon \frac{\partial^2 c}{\partial x^2} = 0,$$

where $\varepsilon = \sigma/Lv$ is a dimensionless parameter, similar to the inverse Reynolds
number. When ε is small, we have an equation with a small parameter in front
of the highest derivative. This is the case when convection is dominating over
diffusion. When $\varepsilon > 0$, the equation belongs to the *parabolic* class of PDEs,

whereas it is *hyperbolic* when $\varepsilon = 0$. In this case, it is not just the order of the equation that changes, the type changes as well.

Let us consider the solution of the initial value problem:

$$\frac{\partial c}{\partial t} + \frac{\partial c}{\partial x} - \varepsilon \frac{\partial^2 c}{\partial x^2} = 0, \ t > 0, \ -\infty < x < \infty,$$

$$c(x, 0) = f(x).$$

For $\varepsilon = 0$, as we have already seen in Section 7.3, the solution is then simply

$$c(x, t) = f(x - t).$$

The initial density profile moves unchanged to the right with speed 1 (speed v in the original variables). This is also reasonable from a purely physical reasoning.

When $\varepsilon > 0$, it is convenient to choose a coordinate system following the flow:

$$x' = x - t,$$

$$t' = \varepsilon t.$$

Then

$$\frac{\partial c}{\partial t} = \frac{\partial c}{\partial t'} \frac{\partial t'}{\partial t} + \frac{\partial c}{\partial x'} \frac{\partial x'}{\partial t} = \varepsilon c_{t'} - c_{x'},$$

$$\frac{\partial c}{\partial x} = \frac{\partial c}{\partial t'} \frac{\partial t'}{\partial x} + \frac{\partial c}{\partial x'} \frac{\partial x'}{\partial x} = c_{x'},$$

$$\frac{\partial^2 c}{\partial x^2} = c_{x'x'}.$$

Thus,

$$c_t + c_x - \varepsilon c_{xx} = \varepsilon c_{t'} - c_{x'} + c_{x'} - \varepsilon c_{x'x'} = 0,$$

or

$$c_{t'} = c_{x'x'},$$

which is the classical parabolic equation. We leave to the reader to show that the function

$$c_f(x', t') = \frac{1}{\sqrt{4\pi t'}} \exp\left(-\frac{x'^2}{4t'}\right) \tag{10.2}$$

is a solution for $t' > 0$. This is the so-called *fundamental solution*. From the formula

$$\int_{-\infty}^{\infty} e^{-x^2} \, \mathrm{d}x = \sqrt{\pi},$$

we find that

$$\int_{-\infty}^{\infty} c_f(x', t') \mathrm{d}x' = 1$$

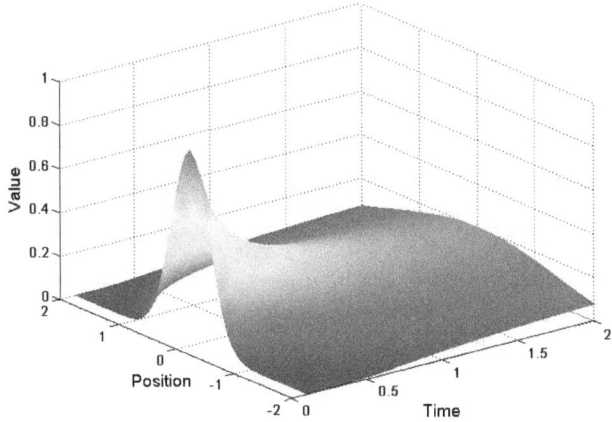

FIGURE 10.2
Time development of the fundamental solution shown for $t = 0.1$ up to 2, and $\varepsilon = 1$.

for all $t' > 0$. The fundamental solution is, in other words, a Gaussian distribution with variance increasing linearly with time. When $t' \to 0$, the solution approaches the δ-function at the origin. Physically, this corresponds to a release of a unit quantity of the substance at the origin at time $t' = 0$. The solution is sketched in Figure 10.2.

If we return to original coordinates, we obtain

$$c_F(x,t) = \frac{1}{\sqrt{4\pi\varepsilon t}} \exp\left(-\frac{(x-t)^2}{4\varepsilon t}\right). \tag{10.3}$$

It is easy to show that the general solution of the initial value problem $c(x,0) = f(x)$ can be expressed as a *convolution integral* with the fundamental solution:

$$c(x,t) = \int_{s=-\infty}^{\infty} f(s)c_F(x-s,t)\mathrm{d}s. \tag{10.4}$$

A general feature of convolution says that the result is at least as nice as the nicest of the functions involved, and actually, the solution $c(x,t)$ will be infinitely many times differentiable for $t > 0$ no matter how f looks. Fast variations are smoothed out more and more as time goes on, and

$$\lim_{t\to\infty} c(x,t) = 0,$$

regardless of x. How fast smearing takes place depends on the size of ε. In this linear equation, the solution becomes smoother and smoother as time passes, but for more general non-linear convection/diffusion equations this does not need to be the case.

On its way towards the final stage, the solution of (10.4) passes what is called an *intermediate asymptotic* state. The concept was introduced in [6] and says (in the simplest case) that the solutions of differential equations may asymptotically approach simplified solutions as time passes, and before they reach their final stage.

Consider the diffusion equation

$$\frac{\partial c^*}{\partial t^*} = \kappa \frac{\partial^2 c^*}{\partial^2 x^*}, \ 0 < t^*, \ -\infty < x^* < \infty,$$

with initial condition $c^*(x^*, 0) = f(x^*)$, and where f is a function localized in the interval $[-L, L]$ such that

$$\int_{-L}^{L} f(x^*) dx^* = Q_0,$$

$$f(x^*) = 0 \text{ for } L \leq |x^*|.$$

For large times, $t = \mathcal{O}(T)$, the solution, due to diffusion, has a spatial extension $X = \mathcal{O}(\sqrt{\kappa T})$, where $X \gg L$ (However, mathematically, the solution is non-vanishing on the whole interval $[-\infty, \infty]$ for all $t > 0$).

Let us scale the problem by T, X and Q_0/X:

$$t^* = Tt,$$

$$x^* = Xx = \sqrt{\kappa T}x,$$

$$c^* = \frac{Q_0}{X} c.$$

This gives

$$\frac{\partial c}{\partial t} = \frac{\partial^2 c}{\partial^2 x},$$

while the initial condition is now

$$f(x) = c(x, 0) = \frac{X}{Q_0} c^*(xX, 0) = \frac{X}{Q_0} f^*(Xx).$$

We see that $f(x) = 0$ for $|x| > \varepsilon = L/X \ll 1$, and

$$\int_{-\infty}^{\infty} f(x) dx = \frac{1}{Q_0} \int_{-\infty}^{\infty} f^*(Xx) X dx = 1.$$

The scaled problem becomes

$$c_t = c_{xx}, \ c(x, 0) = f(x),$$

$$\int_{-\infty}^{\infty} f(x) dx = 1, \ f(x) = 0 \text{ for } |x| > \varepsilon.$$

At $t = O(1)$, the initial condition looks like a δ-function regardless of how irregular f is, and the actual solution must therefore be quite similar to the fundamental solution (the argument can be made mathematically precise by studying the convolution integral in Eq. 10.4). Fundamental solutions therefore have greater applicability than one might think:

For large times, the details of the initial conditions are blurred, and we can use solutions derived from simpler initial conditions.

Linear diffusion equations in one or several dimensions are thoroughly covered in all textbooks on partial differential equations.

10.3   A Nuclear Power Plant Accident

In this case study, we show how one can use conservation principles and basic properties of the fundamental solution of the one-dimensional convection/diffusion equation to analyse a hypothetical release of radioactive material. The analysis is typical of how one will try to get a first rough overview of a relatively difficult modelling problem. The situation is of course fictional, and was presented as an exam task at NTNU in 1986, just after the Chernobyl disaster.

From a nuclear power plant, there is an uncontrolled release of radioactive cooling water to a river past the power plant. The radioactivity is mainly due to a certain short-lived isotope (numerical values below are selected in order to produce reasonably simple numeric answers).

The special thing here is that the radioactive material breaks down. If we have a solution with a concentration c of radioactive material, the concentration decays exponentially with time,

$$c(t) = c(0) \exp(-t/t_0).$$

This could also be described by the differential equation

$$\frac{dc}{dt} = -\frac{c}{t_0}$$

where the time $t_0 \ln 2$ is the *half-life* for the isotope.

We assume that the river stream flows at a mean velocity $U = 0.2\mathrm{m/s}$. In reality, the water velocity varies with the river topography, and the waters are mixed and spread both by turbulence and because the water velocity is not constant over the cross-section of the river. In practice, it is common to model this by a so-called *eddy-diffusivity* (*eddy* = whirl) along the direction of the river, defined by a diffusion coefficient κ_E. The diffusion coefficient will have the same dimension as the molecular diffusion coefficient but will be much larger. Here, we let κ_E be equal to $1\mathrm{m^2/s}$, which is not a completely unreasonable value. We consider the river to be one-dimensional and assume that emissions come from a stationary point source.

When formulating the conservation law, the density of radioactive material c enters as amount per length unit of the river, while the flux will have two contributions, one from the diffusion and one from convection:

$$J(x,t) = c(x,t)U - \kappa \frac{\partial c}{\partial x}(x,t).$$

The decreasing radioactivity can be modelled as a sink with intensity $\frac{c}{t_0}$. The spill is a δ-function source in $x = 0$. The change of radioactive material within an interval $[x_1, x_2]$ of the river is thus described by the conservation law

$$\frac{d}{dt} \int_{x_1}^{x_2} c(x,t)\, dx + J(x_2,t) - J(x_1,t) = \int_{x_1}^{x_2} \left(-\frac{c(x,t)}{t_0} + q(t)\delta(x) \right) dx.$$

The differential form of the equation will be

$$\frac{\partial c}{\partial t} + U \frac{\partial c}{\partial x} - \kappa \frac{\partial^2 c}{\partial x^2} = -\frac{c}{t_0} + q(t)\delta(x),$$

which is a linear convection/diffusion equation with a source/sink term. Assume that the discharge has been going on with a constant amount q_0 per unit of time from time $t = 0$. The total amount of radioactive material in the river at any time can then be calculated from the conservation law by integrating from $x_1 = -\infty$ to $x_2 = \infty$:

$$\frac{dC}{dt} = -\frac{C}{t_0} + q_0,$$

$$C(t) = \int_{-\infty}^{\infty} c(x,t)\, dx.$$

Note that the flux terms disappear since $\lim_{x \to \infty} J(x,t) = \lim_{x \to -\infty} J(x,t) = 0$. The solution for $C(t)$ under the assumption that $C(0) = 0$ follows immediately:

$$C(t) = q_0 t_0 \left(1 - \exp(-t/t_0)\right).$$

As $t \to \infty$, the total amount converges to $q_0 t_0$.

Consider now a situation where a constant discharge q_0 lasts for $t_1 = 30$ minutes. We are seeking an approximate solution for $c(x,t)$ at two different times: (i) immediately after the spill is over, and (ii) after $t_2 = 10^6$s ≈ 11.5 days.

After the 30 minutes, there is a total amount

$$C_0 = t_0 q_0 \left(1 - e^{-t_1/t_0}\right) \approx t_1 q_0$$

in the water (note that $t_1 \ll t_0$). Convection (i.e., the motion of water masses with mean speed U) has led to a spreading of material over a length $L = U t_1 = 360$ m. How much diffusion has affected the solution is estimated by the length scale for diffusion,

$$\sigma_1 = \sqrt{2\kappa t_1} = 60 \text{ m}.$$

This is significantly less than L. Since there is no appreciable radioactive decay during this short period, the concentration, c_0, is nearly constant from $x = 0$ to $x = L$, and mass balance gives $c_0 L = q_0 t_1$. Thus,

$$c(x, t_1) \approx \begin{cases} c_0 = q_0/U, & 0 \le x \le L \\ 0, & \text{otherwise.} \end{cases}$$

Actually, the solution should be a little "rounded" on both ends, but we have a good approximation when neglecting both radioactive decay and turbulent diffusion.

After 10^6 s, the dispersion and decay can no longer be neglected. The length scale for the diffusion is now

$$\sigma_2 = \sqrt{2\kappa t_2} \approx 1400 \text{ m}, \tag{10.5}$$

which is significantly larger than the original length L of the discharge. The total amount of radioactive material is given by

$$\int_{-\infty}^{\infty} c(x, t_2) \, dx = \left[q_0 t_1 (1 - e^{-t_1/t_0}) \right] e^{-(t_2 - t_1)/t_0} \approx q_0 t_1 \exp(-t_2/t_0).$$

Relative to the amount just after the end of the discharge, the remaining amount is about $\exp(-t_2/t_0) \approx 0.8 \cdot 10^{-12}$ less.

Since $t_1 = 30$ minutes is much less than $t_2 = 10^6$ seconds, and $L/\sigma_2 \approx 0.25$, we can, with high accuracy, assume that all emissions of radioactive material occurred at time zero. It is then possible to exploit the fundamental solution to the convection/diffusion equation stated in (10.3), giving the solution of a unit discharge at $t = 0$ and $x = 0$. The full solution becomes, approximately,

$$c(x, t_2) \approx [q_0 t_1 \exp(-t_2/t_0)] \frac{1}{\sqrt{4\pi \kappa t_2}} e^{-(x - U t_2)^2/(4\kappa t_2)},$$

where the first part denotes the total remaining radioactive material and $U t_2 = 200$ km.

It is actually easy to write down the solution for an arbitrary time-variable discharge $q(t)$ from a point-source. Since we can assume that the emissions at different times do not affect each other, it is possible to consider emissions as a series of point-discharges, and then sum up the corresponding solutions. In the same way as above, the solution of a discharge over time duration $d\tau$ at a time $\tau < t$

$$dc(x, t) = [q(\tau) d\tau \exp(-(t - \tau)/t_0)] \, F(x, t - \tau).$$

The distribution of radioactive material at time t is then given by the integral

$$c(x, t) = \int_{-\infty}^{t} dc(x, t) = \int_{-\infty}^{t} [q(\tau) d\tau \exp(-(t - \tau)/t_0)] \, F(x, t - \tau) d\tau.$$

In practice, the radioactive emissions often include several different isotopes. Decomposition of one isotope could also lead to other radioactive isotopes. This will lead to connections between the conservation laws for the individual isotopes, but the link is limited to the source-terms. Moreover, the modelling of the river will naturally also be made considerably more advanced. In particular, turbulent mixing is described by models that relate the strength of the diffusion to the amount of turbulent kinetic energy. The modelling will first calculate the flow and turbulence level in the water using a hydrodynamic turbulence model, and then run a *transport model* calculating the distribution of the radioactive material on the basis of the hydrodynamic solution.

Today, the authorities require that such models are developed and tested *before* an accident occurs (the models can actually be tested by a controlled release of radioactive isotopes).

10.4   Similarity Solutions

So far we have always been able to scale the variables in our equations, but for some "academic" problems there are no natural scales to use. The fundamental solution to the linear diffusion equation is one of such simple examples. It is impossible to find reasonable time and space scales for this problem by just looking at the equation and the definition domain, $-\infty < x < \infty$ and $0 < t$. However, if there are no scales to use, we must *combine* the variables in order to obtain dimensionless equations.

We will not go into further detail on the theory of similarity solutions, where, in particular, the Norwegian mathematician Sophus Lie made important contributions ([6], [31], [36]). However, we will, based on dimensional analysis, illustrate the method by means of an example from heat conduction. The idea of this section has been taken from [15].

Two infinite materials with different but constant temperatures are brought into contact at time $t = 0$, as shown in Figure 10.3. We assume that heat is transported smoothly through the contact surface, and we assume one-dimensional heat conduction. The problem is to determine the temperature development in the material as time goes on. Since we assume infinite extent, and we consider the time from 0 to ∞, there is no length or time scale. In practice, the materials will have a finite extent, L, and after some time T, the temperature at the ends begins to change. As long as we limit ourselves to times that are significantly smaller than T, we should be able to use the solution valid for an infinite extent of the material. Heat transfer and storage of a material is (in its simplest form) determined by three material constants: *mass density*, ρ, $[\rho] = \mathrm{kgm}^{-3}$, *specific heat capacity*, c, $[c] = \mathrm{Jkg}^{-1}\mathrm{K}^{-1}$, and the *heat conduction coefficient*, k, $[k] = \mathrm{Js}^{-1}\,\mathrm{m}^{-1}\mathrm{K}^{-1}$.

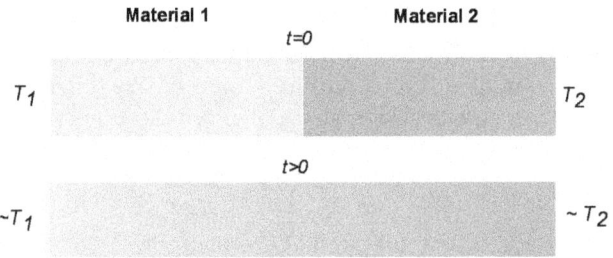

FIGURE 10.3
The temperature in the blocks for $t = 0$ and after some time in contact.

We introduce a dimensionless temperature τ by writing

$$T = T_1 + (T_2 - T_1)\tau,$$

and quite generally we expect

$$\tau = \tau(x, t, \rho_1, c_1, k_1, \rho_2, c_2, k_2), \tag{10.6}$$

where the indices indicate material 1 and 2. If we know the heat conduction equation, we know that ρ, c, and k in the material combines into a heat diffusion coefficient $\kappa = k/c\rho$, while the heat fluxes are generally of the form $-k\nabla T$. The heat flux should be continuous across the border between the materials. This means that we must be able to simplify (10.6) to

$$\tau = \tau(x, t, \kappa_1, \kappa_2, k_1, k_2).$$

The dimension matrix for these six variables has rank 3 (check!), and we have therefore also three dimensionless combinations, for example,

$$\eta = \frac{x}{\sqrt{\kappa_1 t}},$$

$$a = \frac{k_1}{k_2},$$

$$b = \frac{\kappa_1}{\kappa_2}.$$

Thus, we have found that the solution of the problem must be written in the form

$$\tau(x, t) = \tau_0(\eta, a, b) = \tau_0\left(\frac{x}{\sqrt{\kappa_1 t}}, \frac{k_1}{k_2}, \frac{\kappa_1}{\kappa_2}\right).$$

As expected, the solution depends on a combination of x and t, and already here we make a surprising observation:

$$\tau(0, t) = \tau_0(0, a, b)$$

is constant and independent of time for $t > 0$! Unfortunately, dimensional analysis can not give us the exact expression for $\tau(0, t)$, but the expression for τ has reduced the problem from a partial to an ordinary differential equation. The heat conduction equation in material 1 now takes the form

$$\frac{\partial \tau}{\partial t} = \kappa_1 \frac{\partial^2 \tau}{\partial x^2},$$

and by using $\frac{\partial \tau}{\partial t} = \frac{d\tau_0}{d\eta} \frac{\partial \eta}{\partial t}$ etc., we find

$$\frac{d^2 \tau_0}{d\eta^2} + \frac{1}{2} \eta \frac{d\tau_0}{d\eta} = 0,$$

and similarly for material 2,

$$\frac{d^2 \tau_0}{d\eta^2} + \frac{1}{2} b\eta \frac{d\tau_0}{d\eta} = 0. \tag{10.7}$$

Equation (10.7) has general solution

$$\tau_0 = A_2 + B_2 \operatorname{erf}\left(\frac{\sqrt{b}}{2}\eta\right),$$

whereas for material 1 we get

$$\tau_0 = A_1 + B_1 \operatorname{erf}\left(\frac{1}{2}\eta\right).$$

We have 4 constants from the integration, but since $\tau_0(\eta) \to 0$ when $\eta \to -\infty$, and $\tau_0(\eta) \to 1$ when $\eta \to \infty$, only two remain:

$$\tau_0 = A\left(1 + \operatorname{erf}\left(\frac{1}{2}\eta\right)\right), \quad \eta < 0,$$

$$\tau_0 = 1 + B\left(\operatorname{erf}\left(\frac{\sqrt{b}}{2}\eta\right) - 1\right), \quad \eta > 0.$$

The solution found requires that $t > 0$, and then τ_0 is continuous at $\eta = 0$. This gives

$$A = 1 - B.$$

Finally, we use that the flux must be continuous across the contact surface,

$$k_1 \frac{\partial T}{\partial x}\bigg|_{0-} = k_2 \frac{\partial T}{\partial x}\bigg|_{0+}.$$

After introducing τ and η, this gives

$$a \frac{d\tau_0}{d\eta}\bigg|_{0-} = \frac{d\tau_0}{d\eta}\bigg|_{0+},$$

TABLE 10.1
Approximate values of density, specific heat, and heat conduction coefficient for the human foot and burning coal

	ρ [kgm^{-3}]	c[J/kgK]	k[W/mK]	T[°C]	w[m²Ks$^{.5}$]
Foot	1000	4000	0.6	37	1550
Hot coal	150	800	0.04	600	70

or

$$aA\frac{1}{2} = B\frac{\sqrt{b}}{2},$$

and finally,

$$A = \frac{\sqrt{b}}{\sqrt{b}+a}, \quad B = \frac{a}{\sqrt{b}+a}.$$

We get a surprisingly simple expression for the temperature of the interface:

$$T(0,t) = T_1 + (T_2 - T_1)\,A = \frac{T_1 w_1 + T_2 w_2}{w_1 + w_2}, \quad w_i = \sqrt{\rho_i c_i k_i}.$$

The material constant $w = \sqrt{\rho c k}$ is called the *thermal effusivity*.

Now we could stop here, but the expression explains why, in winter, it feels much colder to touch a piece of metal than a piece of wood. In summer, we may get burned by a piece of metal in the sun, whereas touching a piece of wood with the same temperature is without any risk. The explanation is as follows: Your finger will have a temperature T_f and a certain w_f. Although wood and metal have the same temperature T_0, $w_{wood} < w_f$, while $w_f \ll w_{metal}$. This means that $T(0,t) \approx T_f$ when we touch the wood, while $T(0,t) \approx T_0$ when we touch the metal.

This reasoning can actually be taken even further. Some literature search has revealed the values in Table 10.1.

These values give a contact temperature T_k between a *human foot* and *hot coal* given by

$$T_k = \frac{37°\text{C} \times 1550 + 600°\text{C} \times 70}{1550 + 70} \approx 62°\text{C}.$$

By no means deterring! Such an explanation of why it is possible to walk across a hot coal bed is, of course, rejected by a compact majority on the Internet. Some years ago, a Swedish physics professor visited NTNU and let students try this astonishing experiment. He waited until the top of the coal was burned out so that it screened somewhat for exposure to the heat radiation from below. As far as we know, no one was hurt by walking over the bed.

One might wonder for how long time one could trust the similarity solution. If we assume that the outer layer of the skin (*epidermis*) is about $L = 0.5$mm, it is possible to estimate the time by setting $\eta = 1$, or

$$t = \frac{L^2}{\kappa} = \frac{\left(10^{-3}/2\right)^2 \times 1000 \times 4000}{0.6}\text{s} = 1.7\text{s}.$$

It is not worth stopping!

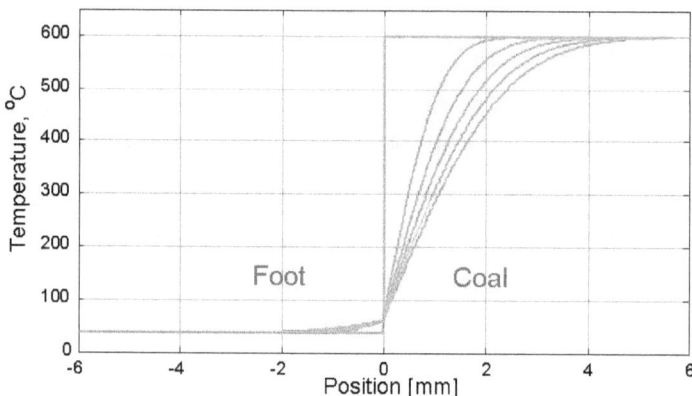

FIGURE 10.4
The temperature profile across the contact for a situation with the foot to the left and hot coals to the right is shown for 0–5 seconds. Note that both the temperature and flux are continuous at $x = 0$ when $t > 0$.

The temperature for this particular example is shown for 0–5 seconds on Figure 10.4. As one can see, our estimate is realistic, but the temperature increase has reached far less than 2 mm into the foot even after 5 seconds.

10.5   Non-linear Diffusion

While models leading to linear diffusion equations are relatively simple to analyse, new and unexpected things occur when the diffusion coefficient depends on the dependent variable and the equation becomes nonlinear.

Gas flowing isothermally in a porous medium is a simple example that leads to a non-linear diffusion equation. A porous medium, such as sandstone, has small pores where gas can flow in. *Porosity*, ϕ, is the volume fraction of pores so that $0 < \phi < 1$. *Darcy's law* says that the volume flow (m^3/m^2 s) is given by

$$\mathbf{q} = -\frac{K}{\mu}\nabla p,$$

where K is a constant of proportionality called the *permeability*. Furthermore, μ is the dynamic viscosity of the gas and p the pressure. We shall assume that the state equation for the gas has the form $p = a\rho^\gamma$, where $\gamma > 0$. The conservation law for the gas will then be

$$\frac{d}{dt} \int_R \phi\rho dV + \int_{\partial R} \rho\mathbf{q} \cdot \mathbf{n} d\sigma = 0,$$

which gives us the differential formulation

$$\phi\frac{\partial\rho}{\partial t} + \nabla \cdot (\rho\mathbf{q}) = 0.$$

If we insert the state equation in the expression for $\mathbf{q}$, we obtain

$$\mathbf{q} = -\frac{K}{\mu}\nabla p = -\frac{Ka\gamma}{\mu}\rho^{\gamma-1}\nabla\rho,$$

and therefore

$$\frac{\partial\rho}{\partial t} = \nabla \cdot \left(\frac{Ka\gamma}{\mu\phi}\rho^\gamma\nabla\rho\right).$$

As we see, we have got a diffusion equation with a diffusion coefficient which is proportional to ρ^γ. In particular, we see that the diffusion coefficient approaches 0 when the density tends to 0 (when γ is greater than 0). There is a lot of theory for non-linear diffusion equations although it is no longer possible to apply the superposition principle for solutions. Even for such equations, there are similarity solutions. For equations

$$\frac{\partial\rho}{\partial t} = \frac{\partial}{\partial x}\left(\kappa(\rho)\frac{\partial\rho}{\partial x}\right)$$

one may look at solutions on the form $\rho(x,t) = g(s)$, $s = x/t^{1/2}$. If we insert this, we obtain an ordinary differential equation for g:

$$\kappa(g)g'' + \kappa'(g)g'^2 + \frac{s}{2}g' = 0.$$

For $\kappa(g) = 1$ the equation is reduced to $g'' + sg'/2 = 0$ with the well-known solution

$$g(s) = A\int_{-\infty}^s e^{-\xi^2/4}d\xi + B.$$

If $\kappa(\rho) = \rho$, we obtain a solution that is sketched in Figure 10.5. The solution is 0 at a finite value of s, and therefore, $x_{max} = s_{max}\sqrt{t} \propto \sqrt{t}$. For this diffusion equation the solution spreads out at a final speed! See [12] for a more detailed analysis of diffusion.

Another class of solutions is the so-called *Barenblatt solutions* for $\kappa = \kappa_0 (\rho/\rho_0)^m$:

$$\rho(x,t) = \rho_0 \left(\frac{t_0}{t}\right)^{1/(m+2)} \left(1 - \left(\frac{x}{x_1}\right)^2\right)^{1/m} , \quad t > 0, |x| \le x_1 = x_0 \left(\frac{t}{t_0}\right)^{1/(m+2)}$$

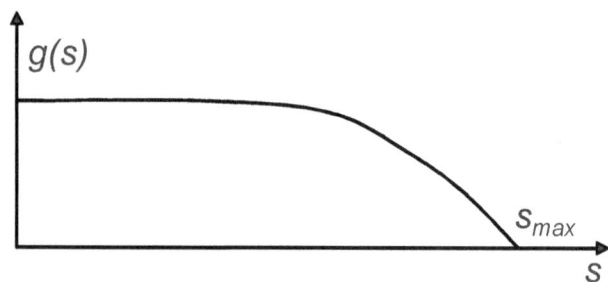

FIGURE 10.5

Sketch of the solution for $\kappa(\rho) = \rho$.

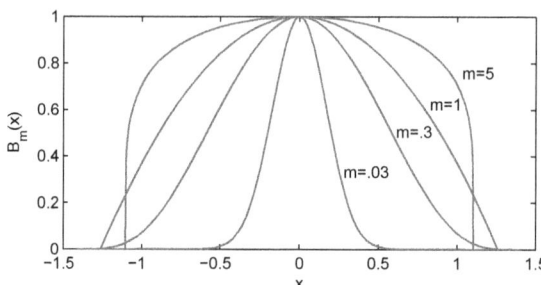

FIGURE 10.6

Barenblatt solutions shown for four different values of m. The solutions are scaled according to their maximum.

(t_0 and x_0 may be expressed by κ_0 and $Q = \int_{x=-\infty}^{\infty} c(x,0)\mathrm{d}x$). Contrary to the standard fundamental solution corresponding to the limit $m \to 0$, the Barenblatt solutions for $m > 0$ have finite extension on the x-axis. Examples of Barenblatt solutions for a selection of m-values are shown on Figure 10.6. Also for these equations there are theorems saying that solutions of random, but localized initial conditions approach the Barenblatt solutions when the time increases. The solutions are described in [20], but not in [6].

10.6   Modelling of Turbulence

The theory of turbulence is a very good example of how we can make use of stochastic considerations in mathematical modelling. We abandon describing the phenomenon in a deterministic way below a certain level, saying that faster variations are *stochastic* or *random*. When a complete description is out of reach, we try instead to model the evolution of the mean values, as

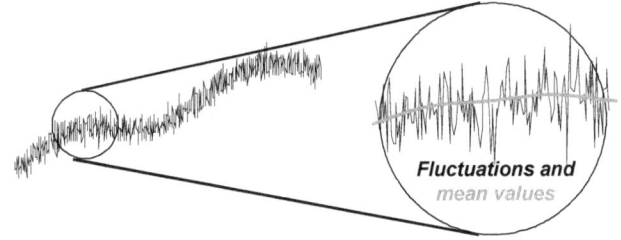

Fluctuations and *mean values*

FIGURE 10.7
An observed signal consisting of rapid fluctuations and a more slowly varying mean.

illustrated in Figure 10.7. We shall illustrate this technique by showing how one obtains equations for turbulence in the very simplest case, namely for an incompressible, viscous fluid without any influence of external forces.

To derive the equations we need the continuity equation and Navier-Stokes equation derived in Chapter 9. We start by the following four differential equations

$$\nabla \cdot \mathbf{v} = 0, \tag{10.8}$$

$$\frac{\partial}{\partial t}\mathbf{v} + \nabla \cdot ((\mathbf{v})\mathbf{v}) = -\frac{1}{\rho}\nabla p + \frac{\mu}{\rho}\nabla^2\mathbf{v}. \tag{10.9}$$

In turbulence theory, a full solution of equations (10.8) and (10.9) is out of reach. The velocity and the pressure are, however, seen as random variables and written as

$$\mathbf{v} = E(\mathbf{v}) + (\mathbf{v} - E(\mathbf{v})),$$
$$p = E(p) + (p - E(p)).$$

Expectation values vary over space and time with scales given by the dimensions of the phenomenon we are considering. These scales are called the *macroscopic scales*. "Small-scale variations" or the so-called *fluctuations* take place at the *microscopic scales*. In turbulence theory it is assumed that these scales are well separated, but in some situations that does not need to be the case.

Unfortunately, the practice of turbulence theory is the opposite of what is used in statistics and probability theory: Deterministic variables are denoted by capital letters and random variables with lower case letters. In order not to confuse readers with a background in mechanics, we shall stick to this practice. Thus, we use $\mathbf{V}$ for $E(\mathbf{v})$, and $\mathbf{v} - E(\mathbf{v})$ by $\mathbf{v}$:

$$\mathbf{v} := \mathbf{V} + \mathbf{v},$$
$$p := P + p. \tag{10.10}$$

Here $\mathbf{V}$ and P are deterministic functions of the macroscopic variables $\mathbf{x}$ and t, while $\mathbf{v}$ and p are stochastic variables with expectations 0, $\mathsf{E}(\mathbf{v}) = 0$, $\mathsf{E}(p) = 0$. Often it is assumed that $\mathbf{v}$ and p are Gaussian variables, and in general, the parameters in the probability distributions of $\mathbf{v}$ and p will depend on $\mathbf{x}$ and t.

We now introduce (10.10) into (10.8) and apply the expectation operator, which we assume commutes with the differentiation:

$$0 = E\left(\nabla \cdot (\mathbf{V} + \mathbf{v})\right) = \nabla \cdot \mathbf{V} + \nabla \cdot E\mathbf{v} = \nabla \cdot \mathbf{V}.$$

Thus we see that $\mathbf{V}$ also satisfies the (10.8)

$$\nabla \cdot \mathbf{V} = 0 \qquad\qquad (10.11)$$

and even $\nabla \cdot \mathbf{v} = 0$.

If (10.9) is treated in the same way, we obtain for component "j"

$$\rho\left(\frac{\partial V_j}{\partial t} + \nabla \cdot (V_j \mathbf{V}) + \nabla \cdot E\left(v_j \mathbf{v}\right)\right) = -\frac{\partial P}{\partial x_j} + \nabla \cdot (\mu \nabla V_j).$$

It is usual here to insert the tensor $\mathbf{T} = -\rho E(\mathbf{v}\mathbf{v}')$, that is, $T_{ij} = -\rho E(v_i v_j)$. This tensor is called the Reynolds stress tensor. The momentum equation for the macroscopic variables is then

$$\rho(\mathbf{V}_t + \nabla \cdot ((\mathbf{V})\mathbf{V})) = -\nabla P + \nabla \cdot (\mu \nabla \mathbf{V} + \mathbf{T}). \qquad (10.12)$$

Equation (10.12) differs from Eq. 10.11 by an extra force term, $\nabla \cdot \mathbf{T}$, whereas (10.11) is essentially the same as (10.8).

By subtracting (10.12) from (10.9) and inserting $\mathbf{V} + \mathbf{v}$ etc., we obtain an equation for $\mathbf{v}$. If we then take the scalar product with $\mathbf{v}$ and apply the expectation operator, the result is a transport equation for *turbulent energy per unit volume*,

$$e = \rho \frac{1}{2} E(\mathbf{v} \cdot \mathbf{v}).$$

The fundamental question now is how to express the Reynolds stresses by means of macroscopic variables. This is the main problem in turbulence theory. The oldest attempt was made by Boussinesq as early as 1877. He defined $\mathbf{T}$ as a function of $\partial V_i / \partial x_j$ in the same way as the stress tensor for a Newtonian fluid. Thus we can write equation

$$\rho(\mathbf{V}_t + \mathbf{V} \cdot \nabla \mathbf{V}) = -\nabla P + \nabla \cdot ((\mu + \mu_T)\nabla \mathbf{V}),$$

and the conservation laws of mass and momentum have exactly the same form as before. The constant μ_T, with the dimension of viscosity, represents an additional "viscosity" due to turbulence, called *eddy viscosity*. It has been shown that this model matches free turbulent flow well, where μ_T can be 10 to 1000 times greater than μ. Since it is easy to apply, this has led to extensive use as an adjustment factor in numerical computations. More advanced models relate μ_T to the local turbulent energy.

If a liquid contains a substance that is mixed with the liquid by turbulent diffusion, we will, in addition to Eq. (10.8) and (10.9) have a convection/diffusion equation for the concentration c of the substance (the "substance" may well be heat content):

$$\frac{\partial c}{\partial t} + \mathbf{v} \cdot \nabla c = \nabla \cdot (\kappa \nabla c),$$

where κ is the diffusion coefficient.

In the same way as for $\mathbf{V}$ and P, it is possible to write the concentration $\mathsf{E}c + (c - \mathsf{E}c) := C + c$, and if this is inserted into the equation, we have

$$\frac{\partial(C + c)}{\partial t} + (\mathbf{V} + \mathbf{v}) \cdot \nabla(C + c) = \nabla \cdot (\kappa \nabla(C + c)).$$

When we then apply the expectation operator and apply $\nabla \cdot \mathbf{v} = 0$, we obtain

$$E(\mathbf{v} \cdot \nabla c) = E(\nabla \cdot (\mathbf{v}c)) - E(c\nabla \cdot \mathbf{v}) = \nabla \cdot E(\mathbf{v}c),$$

and

$$\frac{\partial C}{\partial t} + \mathbf{V} \cdot \nabla C = \nabla \cdot (\kappa \nabla C - E(\mathbf{v}c)).$$

The vector $\mathbf{q}_T = E(\mathbf{v}c)$ is called the turbulent flux. As for the Reynolds stresses, it is difficult to relate this flux to macroscopic variables. The simplest way is, of course, again to assume that $\mathbf{q}_T = -\kappa_T \nabla C$, as this gives the same equation as before with a new diffusion constant $\kappa + \kappa_T$.

10.7 Exercises

1. Consider the scaled diffusion equation

$$\frac{\partial c}{\partial t} = \kappa \frac{\partial^2 c}{\partial x^2}, \quad x \in \mathbb{R}, \ t \geq 0,$$

where $c(x, t)$ is the concentration of a substance and κ is a positive constant.

(a) State the fundamental solution $c_f(x, t)$ which represents a unit discharge in the point $x = 0$ at time $t = 0$. What is κ called?

(b) Show that the total amount, $\int_{-\infty}^{\infty} c_f(x, t) \, dx$, remains equal to 1 for all $t > 0$. It is reasonable to consider $c_f(x, t)$ as a "probability density" on $\mathbb{R}$. What is in this case the *mean value*, η, and *standard deviation*, σ?

(c) In statistics $\pm 3 \times \sigma$ is often said to be the typical extension of a distribution. What is the $[-3\sigma, 3\sigma]$-interval here?

(d) The equation

$$\frac{\partial c}{\partial t} + v\frac{\partial c}{\partial x} = \kappa\frac{\partial^2 c}{\partial x^2},$$

where v is a positive constant, represents a situation where what we are considering is moving to the right with constant speed v. What is the fundamental solution in this case?

(e) Show that if the distribution at $t = t_0$ is $c(x, t_0) = h(x)$, the general solution of the equation for $t > t_0$ may be written

$$c(x, t) = h * c_f(\cdot, t - t_0)(x, t) = \int_{-\infty}^{\infty} h(s) c_f(x - s, t - t_0)\, ds.$$

(f) Use the expression in **(e)** (or a smarter way, based on the uniqueness of solutions) to find the solution for $t = 3$ when the solution of $t = 1$ is

$$h(x) = c_f(x, 1).$$

Can you derive a general property of convolutions of the fundamental solutions based on this?

(g) Which condition on the diffusive flux $\mathbf{j}_d$ must hold at x_0 if there is a dense wall there, so that all diffusion occurs to the right of the wall $(x \geq x_0)$.

(h) Show (or argue) that if x_0 in **(g)** is larger than 0, it is possible to write the solution for a unit emission at $x = 0$ for $t = 0$ as $c(x, t) = c_f(x, t) + c_f(x - 2x_0, t)$, $x \geq x_0, t > 0$.

(i) Determine the fundamental solutions for

$$\frac{\partial c}{\partial t} = \kappa\left(\frac{\partial^2 c}{\partial x^2} + \frac{\partial^2 c}{\partial y^2}\right), \quad (x, y) \in \mathbb{R}^2, \ t \geq 0,$$

$$\frac{\partial c}{\partial t} = \kappa\left(\frac{\partial^2 c}{\partial x^2} + \frac{\partial^2 c}{\partial y^2} + \frac{\partial^2 c}{\partial z^2}\right), \quad (x, y, z) \in \mathbb{R}^3, \ t \geq 0.$$

As long as we are interested in solutions defined on $\mathbb{R}^n$, we may use the fundamental solution to express more general solutions. The two-dimensional equation above may, e.g., describe the extent of an oil spill on the surface of the sea.

(j) Determine how a unit emission at $t = 0$ from an off-shore oil platform situated at $x = 0$ will evolve if there is, in addition, a constant current in the sea. A variable emission over a period of time can be approximated as a number of small spills occurring at constant time intervals. State an expression for the solution in this case. The extent of the oil slick could be defined as the area where the oil film is thicker than a certain thickness, e.g., 5 molecular diameters. *Challenge*: Program and visualize this slick using Matlab or Octave.

2. A chemical dissolved in a liquid spreads by molecular diffusion, modelled by means of the flux

$$\mathbf{j} = -\kappa \nabla \varphi,$$

where φ is the concentration of the substance (kg/m^3), κ is a constant, and ∇ the gradient. What is the unit of κ? The substance decays over time, and during the time from 0 to t, the concentration (if nothing else happens) decreased as

$$\varphi(\mathbf{x},t) = \varphi_0 e^{-t/t_0}.$$

This represents a distributed sink, but how can this be described as a function $q(\mathbf{x},t)$? Use the general conservation law to derive the differential equation

$$\frac{\partial \varphi}{\partial t} = \kappa \nabla^2 \varphi - \frac{\varphi}{t_0}.$$

11

Modelling Projects

This chapter contains a selection of eight small modelling projects which may be attached by the techniques we have been through in the book. A modelling project differs from regular exercises by being rather open and may sometimes have no real solution. In fact, it is often said that modelling projects are most instructive if they address topics that have no complete solution. The projects are presented without any solutions, leaving it to the student to judge when enough is enough.

11.1   The Student 10 km Race

Many years ago, the students' sports club at NTNU arranged the *Student 10 km Race* at the old Trondheim Stadium. The arrangement was immensely crowded with more than 1000 participants, all starting (or trying to start) at the same time. This spectacular event, often in rain and on an extremely dirty track, is the origin of the following case study.

Consider an ordinary race-track with length $L = 400$m. We assume that the mean running speed v^* decreases linearly with the density ρ^*, so that $v^* = v^*_{max}$ for $\rho^* = 0$ students/m, and is $v^* = 0$ m/s when $\rho^* = \rho^*_{max}$. We also consider the track to be one-dimensional and 0-shaped.

(a) State the conservation law for students (assuming no late entries or drop-outs) under these conditions, introduce dimensionless variables, and show that the differential formulation for the students' density may be written as

$$\rho_t + (1 - 2\rho)\,\rho_x = 0, \qquad (11.1)$$

where $0 \leq \rho \leq 1$ and $\rho(x, t)$ is a 2π-periodic function of x.

(b) Find (in implicit form) the exact solution of (11.1) when

$$\rho(x, 0) = \rho_0 + \varepsilon \cos(x), \quad (0 < \varepsilon < \rho_0).$$

Sketch the characteristics for the solution in the xt-plane and show that, as a solution of the integral conservation law, it breaks down and forms a shock in the density after some time.

DOI: 10.1201/9781003725206-11

(c) When will the shock start, and what happens to the shock when $t \to \infty$?

Hint: Start by determining the crossing points for characteristics starting at $3\pi/2 - \theta$ and $3\pi/2 + \theta$, when θ varies from 0 to π. Try to prove that the crossing points lie on a straight line segment, and that there are no crossings elsewhere. Finally, check that the line segment is a shock and that the corresponding solution indeed satisfies the integral conservation law.

11.2 Two Phase Porous Media Flow

Oil is found in porous rocks. Often the porous rock is trapped between layers of solid impermeable rock, called an *oil reservoir*, and when we pump oil out, ground water enters. In so-called *enhanced oil recovery*, water is actively pumped into the reservoir and the oil is forced out. The simultaneous flow of oil and water is complex, and in order to study this in detail, small samples of rock are taken from the reservoir and investigated in the laboratory. The following model is essential in these investigations.

A long thin cylinder of porous sandstone with constant cross-section A is situated along the x-axis. The pores occupy a constant fraction Φ of the volume $(0 < \Phi < 1)$, and are initially filled with oil. We assume that the oil has a constant density and measure the amount of oil by its volume. The sides of the cylinder are closed, but by applying a pressure at one end, it is possible to press oil or water through the stone.

In order to find an expression for the flux of oil in the x-direction, j [m^3/(m^{2}s)], we assume it only depends on the *viscosity*, μ [kg/ms], the *permeability* (inverse flow resistance) of the stone, K [m^2], and the *pressure gradient*, $\partial p/\partial x$.

(a) Show that dimensional analysis gives

$$j = -k \frac{K}{\mu} \frac{\partial p}{\partial x},$$

where k is a dimensionless constant.

Assume that the pores of the cylinder in addition to oil also contains water. All pores are either filled with water or oil, so that a volume V of rock contains a volume $S_o \Phi V$ of oil and $S_w \Phi V$ of water, where $S_o + S_w = 1$. We assume that water and oil have the same pressure and that the corresponding fluxes may be written as

$$j_i = -k_i \left(S_i\right) \frac{K}{\mu_i} \frac{\partial p}{\partial x}, \quad i = o, w.$$

The parameter $k_i \left(S_i\right)$ is called the relative permeability.

(b) Establish the conservation laws for oil and water for the part of the cylinder between $x = a$ and $x = b$. Show that if we apply a pressure gradient such that

$$q = j_o + j_w = \text{constant},$$

then we have, for $S \equiv S_w$, the following hyperbolic equation for S:

$$\Phi \frac{\partial S}{\partial t} + \frac{\partial}{\partial x} \left(\frac{qk_v(S)/\mu_v}{k_o \left(1 - S\right)/\mu_o + k_v \left(S\right)/\mu_v} \right) = 0.$$

(c) Assume that $\mu_o = \mu_w$, $k_o \left(1 - S\right) = 1 - S^2$ and $k_v(S) = S^2$. Solve the equation (11.2) for $t > 0$ for a cylinder of length L when

$$S(x,0) = 1 - x/L, \ 0 \le x \le L,$$
$$S(0,t) = 1, \ 0 \le t.$$

11.3 Reduced Speed Limit

In this problem, we are considering the *standard model* for the traffic of cars along a one-way road.

(a) Describe the basis of the standard model. State the hyperbolic equation the model leads to (when no cars are assumed to enter or leave the road). When will the car density develop shocks?

(b) Between $x = 0$ and $x = 1$ there is now a reduction in the speed limit such that the maximal speed reduces to $1/2$, while the maximum density remains the same. We assume that a similar linear relation between the car velocity and the density also applies for this part of the road.

Which condition on the flux of cars has to hold at $x = 0$ and $x = 1$? Find the solution $\rho(x,t)$ for $t > 0$ and all x when

$$\rho(x,0) = \begin{cases} 1/2, & x < 1, \\ 0, & x > 1. \end{cases}$$

Hint: The density ρ between 0 and 1 remains constant for all $t \ge 0$.

11.4 Traffic Lights at a Pedestrian Crossing

In this problem we study the traffic along a one-way street, and without cars entering or exiting in the first part. All variables are scaled so that the car density ρ is between 0 and 1, and car velocity v is equal $1 - \rho$.

(a) Show how to find an expression for the shock velocity U of a jump in car density, and derive that in this case, $U = 1 - \rho_1 - \rho_2$, where ρ_1 and ρ_2 are the densities on each side of the shock.

For $t < 0$, there is a constant car density $\rho = 1/2$ on the street. Between $t = 0$ and $t = 1$ the cars face a *red* light at a pedestrian crossing at $x = 0$. For $t > 1$, the light is again *green*.

(b) Determine the solution $\rho(x, t)$ for $t \geq 0$.

(*Hint*: Make a sketch of the situation in an x/t-diagram. Show that the solution for ρ has to be found in five different domains, of which the values in four of them are obvious. In order to determine the domains it is necessary to determine their exact borders).

At another place on the street, a second one-way street of the same type as the first merges with the first street.

(c) Which condition must hold at the junction? Assume that the flux on the first street towards the junction is constant, $j_1 = 1/8$, and the corresponding car density is less than $1/2$. Describe the development of car density on the streets when the density ρ_2 on the second street increases from 0 to 1. The drivers on the first street have the right of way, but are flexible and let cars enter from the second street if this is possible. In particular, look at what happens when the flux on the second street reaches $1/8$.

11.5 A Water Cleaning System

A part of a water cleaning system is modelled as a tube of length L along the x^*-axis, where polluted water flows with constant velocity V. The tube also contains absorbers that remove the pollution. The concentration of pollutant in the water is c^*, measured as the amount per length unit of pipe. Similarly, the amount of absorbed pollutant per length unit of pipe is denoted ρ^*. The maximum value of ρ^* is A. Some of the absorbed pollutant will over time re-enter the water stream. The absorption and re-entering is modelled by the equation

$$\frac{\partial \rho^*}{\partial t^*} = k_1 (A - \rho^*) c^* - k_2 \rho^*.$$

(a) State the integral conservation law for the pollutant and show that this leads to the differential form

$$\frac{\partial}{\partial t^*} (c^* + \rho^*) + \frac{\partial}{\partial x^*} (V c^*) = 0.$$

Based on the integral law, establish that a discontinuity in the concentrations, moving with velocity U^*, has to fulfill

$$U^* = \frac{c_2^* - c_1^*}{(c_2^* + \rho_2^*) - (c_1^* + \rho_1^*)} V,$$

where (c_1^*, ρ_1^*) and (c_2^*, ρ_2^*) are the concentrations on each side of the discontinuity.

(b) Introduce suitable scales and show that the differential equations may be written

$$\frac{\partial}{\partial t}(c + \rho) + \frac{\partial c}{\partial x} = 0, \tag{11.2}$$

$$\varepsilon \frac{\partial \rho}{\partial t} = (1 - \rho)c - \beta\rho. \tag{11.3}$$

Explain the meaning of ε and β (*Hint*: Use the same scale for ρ^* and c^*).

Assume that the tube is infinitely long in both directions and consider analytic solutions of equations (11.2) and (11.3) in the form of "fronts", travelling with velocity U, that is,

$$c(x, t) = C(\eta), \quad \rho(x, t) = R(\eta), \tag{11.4}$$

with $\eta = x - Ut$. We limit ourselves to the special case where $C(\eta)$ and $R(\eta)$ satisfy

$$\lim_{\eta \to -\infty} C(\eta) = 1,$$

$$\lim_{\eta \to \infty} C(\eta) = 0,$$

$$\lim_{\eta \to -\infty} R(\eta) = \frac{1}{1 + \beta},$$

$$\lim_{\eta \to \infty} R(\eta) = 0.$$

(c) Insert (11.4) into (11.2), integrate once, and use the behaviour at $-\infty$ and ∞ to determine U and a simple relation between C and R. Use this information and (11.3) to determine $C(\eta)$ and $R(\eta)$. How is the behaviour of the solution when $\varepsilon \to 0$? (*Hint*: The equation

$$\frac{dy}{d\zeta} = y\left(-1 + \frac{y}{M}\right)$$

has a solution

$$y(\zeta) = M\frac{1}{1 + \exp\zeta}$$

for $0 < y < M$).

(d) Assume $\varepsilon = 0$ in (11.3) so that the system (11.2–11.3) simplifies to

$$\rho = \frac{c}{c + \beta}, \quad \frac{\partial}{\partial t}\left(c + \frac{c}{c + \beta}\right) + \frac{\partial c}{\partial x} = 0, \quad -\infty < x < \infty, \ t \geq 0.$$

Consider the initial condition

$$c(x, 0) = \begin{cases} 1, & x < 0, \\ 0, & x > 0. \end{cases} \quad -\infty < x < \infty, \ t \geq 0. \tag{11.5}$$

Show that the corresponding solution associated to (11.5) develops a shock. Determine the shock velocity from the expression in point (a) and compare it to the result in (c).

11.6   River Contamination

Contaminants discharged into a river will be transported with the flow (*convection*) and spread due to turbulence mixing and varying water velocity (*diffusion*). Consider a one-dimensional river with mean flow U and the diffusion coefficient κ.

(**a**) Derive the expression for the flux of contaminants under these simple conditions, and find a length scale of the extent of an instantaneous point discharge after this has been carried a length L down the river by means of the current. At the point $x = 0$ there is a continuous discharge of a substance A so that the concentration in the river becomes $a\,(x, t)$. The substance A is converted into substance B with constant rate μ. Thus, for a water sample from the river, we would have

$$\frac{da}{dt} = -\mu a. \quad -\infty < x < \infty,\ t \geq 0. \tag{11.6}$$

The substance B decays with rate λ, and for the same water sample, the concentration $b\,(x, t)$ of B fulfills

$$\frac{db}{dt} = +\mu a - \lambda b.$$

(**b**) State the conservation laws for A and B in the integral and differential forms.

(**c**) The discharge at $x = 0$ takes place at a constant rate q_0 (amount per time unit). Neglect diffusion and decide how far down the river the concentration of substance A is at its highest when we assume that $\lambda = \mu$.

Hint: The differential equation $\frac{dy}{dt} + ky = e^{-kt}$ has the general solution $y(t) = C_1 e^{-kt} + t e^{-kt}$.

11.7   Lake Sedimentation

A river flows into a lake. The river brings sand and clay so that the lake is filled up over time. We shall formulate and analyse a simple one-dimensional model for how the lake is filled, and assume that it reaches from $x = 0$ to $+\infty$, and has a constant depth h at $t = 0$. Conditions across (in the y-direction) are assumed to be constant.

The amount of sand and clay which settle on the bottom per time and area unit is $q\,(x, t)$. We write the depth $z = b\,(x, t)$, $x \geq 0$, $t \geq 0$, and assume $b\,(x, t) \leq 0$. If the bottom tilts (is not horizontal), the particles on the bottom will continue to move, and it has been found that the mass flux is proportional to the slope, that is, the volume flux may be written

$$j = -k \frac{\partial b}{\partial x}.$$

(a) Write the conservation equation in integral form for a part of the bottom , $x_0 \leq x \leq x_1$, and show that the differential form is identical to the heat diffusion equation,

$$\frac{\partial b}{\partial t} = k \frac{\partial^2 b}{\partial x^2} + q. \tag{11.7}$$

(b) Assume that *all* sand and clay enter at $x = 0$ (i.e., $q = 0$ for $x > 0$), and that the amount entering is always sufficient for (11.7) to hold for $t > 0$. Argue that the solution to (11.7) will be a similarity solution in this case, and find $b(x,t)$ for $x \geq 0$ and $t > 0$. (*Hint:* The equation

$$\frac{d^2 y}{d\eta^2} + \frac{\eta}{2} \frac{dy}{d\eta} = 0$$

has the general solution $A + B \, \text{erf} \, (\eta/2)$, where $\text{erf} \, (x) = \frac{2}{\sqrt{\pi}} \int_0^x \exp\left(-s^2\right) ds$).

(c) A more realistic scenario is that the shore, $s(t)$, moves forward into the lake over time. Assume that a constant volume of sand and clay enters the basin per time unit, q_0, and that all sand and clay enter at the shore.

The solution will then have a stationary shape and may be written by means of a function b_0 so that

$$b(x,t) = \begin{cases} 0, & x \leq s(t) = Ut + x_0, \\ b_0(x - Ut - x_0), & x > Ut + x_0. \end{cases}$$

Determine the velocity U and the solution in this case.

11.8   The Insect Swarm

Flying insects sometimes form dense swarms where the insects are attracted to each other. On the other hand, the swarm has a certain extension, which implies that there is also something preventing the insects from coming too close to each other. This modelling study tries to explain this as a balance between the attraction towards the swarm and a random motion modelled as a diffusion.

The model is for simplicity one-dimensional, where the insects are assumed to stay in a straight tube. For more information, see [20] pp. 188–189.

An insect swarm with density $\rho(x^*, t^*)$ is situated in a long tube parallel to the x^*-axis. In the swarm *random flight* (diffusion) contributes to spreading the insects, while the insects in the swarm are also attracted towards the centre of the swarm. This latter effect can be modelled as a mean drift velocity w,

$$w(x^*, t^*) = -K\left(\int_{-\infty}^{x^*} \rho(s^*, t^*)\, ds^* - \int_{x^*}^{\infty} \rho(s^*, t^*)\, ds^* \right).$$

We shall assume that the total amount of insects,

$$M = \int_{-\infty}^{\infty} \rho(x^*, t^*)\, dx^*,$$

remains constant.

(a) Explain why the model for w is not unreasonable, and state the conservation law for insects in integral form.

(b) Introduce the cumulative distribution of insects,

$$v^*(x^*, t^*) = \int_{-\infty}^{x^*} \rho(s^*, t^*)\, ds^*,$$

and show from the conservation law that v^* satisfies the equation

$$\frac{\partial v^*}{\partial t^*} = \sigma \frac{\partial^2 v^*}{\partial x^{*2}} - K(M - 2v^*)\frac{\partial v^*}{\partial x^*}$$

where σ is the diffusion coefficient. We assume that σ and K are constant.

(c) For a swarm with a diameter L there are two characteristic time scales,

$$T_K = \frac{L}{KM} \quad \text{and} \quad T_D = \frac{L^2}{\sigma}.$$

What do these scales signify? Scale the equation for v^* when $x^* = \mathcal{O}(L)$, $t^* = \mathcal{O}(T_K)$, and $T_D \gg T_K$, and show that it can be stated as

$$\frac{\partial v}{\partial t} = \varepsilon \frac{\partial^2 v}{\partial x^2} - (1 - 2v)\frac{\partial v}{\partial x}.$$

What is the interpretation of ε?

(d) Determine $\rho^*(x^*, t^*)$ when

$$\rho^*(x^*, 0) = \begin{cases} \frac{M}{2L}, & |x^*| < L \\ 0, & \text{otherwise} \end{cases},$$

if we ignore the effect of diffusion. How do you expect, in rough terms, that the exact solution looks when diffusion is included?

(e) To examine the solution in (d) after a long time, it is reasonable to consider a length scale $L' = \sigma/(KM)$ and a time scale $T \gg T_K$ and T_D. Show that with this scaling the equation for v is independent of t to leading order. Verify that the leading order equation has a solution

$$\rho^*(x^*, t^*) = A \frac{1}{\cosh^2(Bx^*)}$$

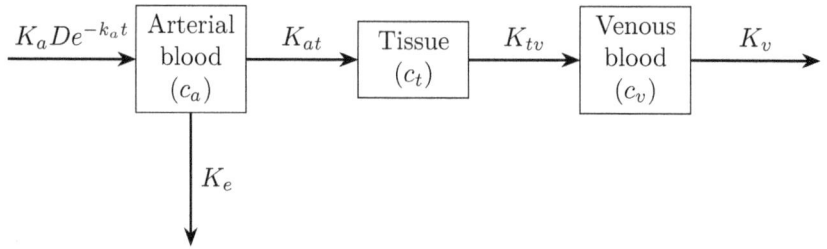

FIGURE 11.1
Flow diagram of the three-compartment open model for drug flow.

and determine A and B for a solution with the initial distribution as
in (**d**).

Hint: Use that $\frac{d}{dx}\tanh(x) = \cosh^{-2}(x)$, and try a v with the correct behaviour when $x \to -\infty$ and $x \to \infty$.

11.9 Drug Distribution after Extravascular Administration

Blood flow in the cardiovascular system is unidirectional. The intravenous
infusion of a drug is absorbed and distributed through arterial blood, tissue,
and venous blood to hit the target. The consumption of the drug from arterial
blood to tissue, tissue to venous blood, and venous blood to the target, flows
at rates k_{at}, k_{tv}, and k_v, respectively. The drug can be cleared from the blood
at a rate k_e. The flow diagram describing the distribution and elimination of
drugs after extravascular administration can be seen in Figure 11.1. We note
that c_a, c_t, and c_v are the drug concentrations in the arterial blood, tissue, and
venous blood, respectively. The term $k_a D e^{-k_a t}$ represents the administration
of the drug.

1. Formulate the mathematical model associated with the flow in Figure 11.1.

2. Show that the above system can be written in the compact form

$$C'(t) = AC(t) + B(t,)$$

   where $A \in M_{n \times n}(\mathbb{R})$ and $B, C : \mathbb{R} \to \mathbb{R}^3$ are to be specified.

3. Provide the general expression of $C(t)$ in terms of $C(0)$.

4. Compute e^{tA} when $k_e = 0$.

5. Assume that $k = k_{at} = k_{tv} = k_v = k_e$.

(a) Show that there exists an invertible matrix P and an upper triangular matrix $\tilde{A}$ such that $A = P\tilde{A}P^{-1}$.

(b) Deduce e^{tA}.

(c) Determine the expression of the solution of the system assuming that $c_a(0) = c_t(0) = c_v = 0(0) = 0$.

12

First-order Quasilinear PDEs

This chapter gives a short introduction to the solution of first-order partial differential equations (PDEs) that occur in connection with models based on conservation principles. It aims at students only with Calculus background. So, what does the title mean? The title contains several words which may be unknown:

- *First Order* = only first-order derivatives occur in the equation.

- *Quasi-linear* = the equation is linear in the highest (first) order derivatives.

- *Partial* = there is more than one independent variable

12.1 Equations and Solutions

The theory below is illustrated for one variable z dependent on two independent variables x and y. Equations with more independent variables are solved in a similar way.

PDEs are divided into several classes, and for an equation to belong to the class in the title, it is necessary that it can be put into what we call the *normal form*. This means that the equation can be written as

$$P\left(x,y,z\right)\frac{\partial z}{\partial x}+Q\left(x,y,z\right)\frac{\partial z}{\partial y}-R\left(x,y,z\right)=0. \tag{12.1}$$

Here P, Q and R are functions only of x, y, and z, and do not contain any derivatives. Note that $\partial z/\partial x$ and $\partial z/\partial y$ only occur in the first power, but there is no such limitation for z in P, Q and R. The reason for the minus in front of the third term will become clear below.

A *solution* of (12.1) is a function

$$z = f\left(x,y\right),$$

which satisfies the equation:

$$P\left(x,y,f\left(x,y\right)\right)\frac{\partial f\left(x,y\right)}{\partial x}+Q\left(x,y,f\left(x,y\right)\right)\frac{\partial f\left(x,y\right)}{\partial y}-R\left(x,y,f\left(x,y\right)\right)=0.$$

DOI: 10.1201/9781003725206-12

If we consider a regular coordinate system, $(x, y, z) \in \mathbb{R}^3$, the function $z = f(x, y)$ will define a *surface* in $\mathbb{R}^3$. Typically, finding a solution to (12.1) means to find a function $f(x, y)$ fulfilling some additional conditions, meaning, having given values on some curve in the xy-plane. It will soon become clear that solving a PDE is radically different from solving ordinary differential equations, although ordinary equations sometimes come up during the solution process.

From Calculus we remember that the vector

$$\mathbf{n} = \left[\frac{\partial f(x, y)}{\partial x}, \frac{\partial f(x, y)}{\partial y}, -1 \right]$$

is a *normal vector* (perpendicular) to the surface $z = f(x, y)$ at the point (x, y, z) (try to derive this yourself if you do not know it).

A *vector field* $\mathbf{V}(x, y, z)$ in $\mathbb{R}^3$ is defined in terms of three functions making up the three components of the vector, say

$$\mathbf{V}(x, y, z) = [P(x, y, z), Q(x, y, z), R(x, y, z)]. \tag{12.2}$$

A vector field defines a set of *stream lines* in space. Curves in space may be parametrized by a variable s and written as

$$\mathbf{r}(s) = [x(s), y(s), z(s)], s \in \mathbb{R}.$$

The stream lines for the vector field $\mathbf{V}$ satisfy the following system of differential equations:

$$\frac{d\mathbf{r}}{ds} = \mathbf{V}(x, y, z), \tag{12.3}$$

or, written out,

$$\frac{dx}{ds} = P(x, y, z),$$

$$\frac{dy}{ds} = Q(x, y, z), \tag{12.4}$$

$$\frac{dz}{ds} = R(x, y, z).$$

In general, one can set $\mathbf{r}(s_0) = \mathbf{r}_0 = [x_0, y_0, z_0]$ and solve the system in (12.4) in order to find the stream line through $\mathbf{r}_0$. In the PDE literature, you often find (12.4) written as

$$\frac{dx}{P} = \frac{dy}{Q} = \frac{dz}{R}.$$

This means *exactly the same* and is nothing but a short way of writing (12.4).

We now make an interesting observation:

- *The normal vector to a solution of (12.1) is perpendicular to the stream lines of the vector field* $\mathbf{V}$, *defined as in 12.2 with P, Q, and R from* (12.1).

This is quite obvious since

$$\mathbf{V} \cdot \mathbf{n} = [P, Q, R] \cdot \left[\frac{\partial f}{\partial x}, \frac{\partial f}{\partial y}, -1 \right] = P\frac{\partial f}{\partial x} + Q\frac{\partial f}{\partial y} - R = 0! \qquad (12.5)$$

In the PDE theory, the stream lines are called *characteristic curves*, or simply *characteristics*.

If $\phi(x, y, z)$ and $\psi(x, y, z)$ are two independent *first integrals* of the quasilinear PDE (12.1), that is, constants along the characteristics, then the general solution of (12.1) is

$$F(\phi, \psi) = 0 \quad \text{or} \quad \phi = G(\psi),$$

where F and G are arbitrary functions.

For example, we consider the quasilinear equation

$$x\frac{\partial z}{\partial x} + y\frac{\partial z}{\partial y} - 1 = 0$$

taken from Volume III of the classic calculus textbook "*Læreboki matematisk analyse*" by R. Tambs Lyche, S 282. The characteristic equations are

$$\frac{dx}{x} = \frac{dy}{y} = \frac{dz}{1}.$$

These results are related to the independent first integrals

$$\phi = \frac{y}{x} = constant, \quad \psi = z - \ln x = constant.$$

The general solution is

$$z = \ln x + G\left(\frac{y}{x}\right), \quad x > 0$$

where G is an arbitrary function.

12.2   Cauchy Problem

We consider the problem:

Given a curve Γ in space. Find a function $z = f(x, y)$ that satisfies (12.1) and is such that the curve Γ is contained in the surface defined by the solution.

The Cauchy problem is the common name for such problems, resembling what we would call an initial value problem for ordinary differential equations.

If we think in practical terms, it is actually not so difficult to imagine how this could be carried out: For all points on Γ, we find the characteristic curves through the points. When we then move along Γ, the characteristics slice out a surface in space. By the way the surface is made, the normal vectors

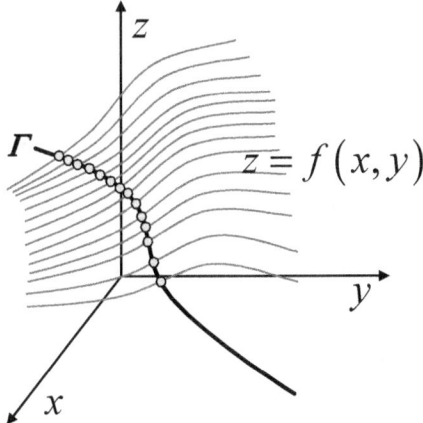

FIGURE 12.1
The characteristics through Γ define a solution to the Cauchy problem.

to the surface must be orthogonal to the characteristics. In other words, we have actually got the situation in (12.5), and have found a solution to (12.1), as illustrated in Figure 12.1. When P, Q, and R are nice and reasonable functions, the solution of (3.4) will be unique. This means that only one characteristic curve can pass through each point in space, or that two different characteristic curves in space can never collide. However, this does not prevent a solution, as the one in Figure 12.1, from "folding over", meaning that one has two different z-values for each point (x, y). We shall see later that this complicates matters for real-world problems where z has a physical meaning.

There are also two special situations where the characteristic method needs to be modified. The first situation is when Γ itself (or parts of it) is a characteristic curve. When this is the case, a *unique* solution to the equation can not be obtained. The second possibility is when there exist functions $f(x, y)$ so that

$$P(x, y, f(x, y)) = 0,$$
$$Q(x, y, f(x, y)) = 0,$$
$$R(x, y, f(x, y)) = 0.$$

Such functions are called *singular solutions*, since they obviously satisfy (12.1).

Even if this way of solving the equations may seem simple, it is quite implicit and not always so easy to carry out. Finally, even if both Γ and the characteristics are known, it may be difficult or even impossible to write the solution in the explicit form $z = f(x, y)$.

Assume that the curve Γ can be parametrized by

$$x = x_0(t), \quad y = y_0(t), \quad z = z_0(t), \quad t \in I. \tag{12.6}$$

Let $\mathbf{r_0} = (x_0, y_0, z_0) \in \Gamma$ be a point corresponding to $t = t_0 \in I$, and let Ω be a domain in $\mathbb{R}^3$ containing $\mathbf{r_0}$.

Uniqueness: *If P, R and R are of class C^1 in Ω, and if*

$$P(\mathbf{r_0})\frac{dy_0(t_0)}{dt} - Q(\mathbf{r_0})\frac{dx_0(t_0)}{dt} \neq 0, \tag{12.7}$$

then there exists in a neighborhood $\mathcal{U}$ of (x_0, y_0) a unique solution of the quasilinear equation (12.1) verifying the initial condition (12.6).

The proof of this result can be found in the textbook written by Zachmanoglou and Thoe [42]. The condition in (12.7) simply says that the initial curve Γ must not be a characteristic.

We consider again the equation

$$x\frac{\partial z}{\partial x} + y\frac{\partial z}{\partial y} - 1 = 0 \tag{12.8}$$

passing through the space curve

$$\mathbf{r_0}(t) = -\mathbf{i} + 2t\mathbf{j} + t^2\mathbf{k}, \ t \in \mathbb{R}, \tag{12.9}$$

where $\mathbf{i}$, $\mathbf{j}$, and $\mathbf{k}$ are unit vectors along the respective axes. We note that $P(x, y, z) = x$, $Q(x, y, z) = y$ and $R(x, y, z) = 1$ are of class C^1. Since

$$P(\mathbf{r_0})\frac{dy_0(t_0)}{dt} - Q(\mathbf{r_0})\frac{dx_0(t_0)}{dt} = -1(2) - 2t(0) = -2 \neq 0,$$

the Cauchy problem (12.8)–(12.9) has a unique solution. The characteristic equations are

$$\frac{dx}{ds} = P(x, y, z) = x,$$

$$\frac{dy}{ds} = Q(x, y, z) = y,$$

$$\frac{dz}{ds} = R(x, y, z) = 1.$$

It is not difficult to solve these equations since they do not interfere with each other:

$$x(s) = C_1 e^s,$$
$$y(s) = C_2 e^s, \tag{12.10}$$
$$z(s) = s + C_3.$$

It remains to find expressions for the special characteristics that pass through the space curve, and thus eliminate the free constants C_1, C_2, and C_3 in (12.10). There are several ways to proceed, and the following is somewhat simpler than the method used in the reference. Let us (without loss of generality) assume that the characteristics cross the curve at $s = 0$. We then obtain

$$x\,(0) = C_1 = -1,$$
$$y\,(0) = C_2 = 2t,$$
$$z\,(0) = C_3 = t^2.$$

The solution is thus the surface defined in *parametric* form as

$$x = C_1 e^s = -e^s,$$
$$y = C_2 e^s = 2t e^s, \qquad\qquad (12.11)$$
$$z = s + C_3 = s + t^2,$$

for the pair of parameters $(s, t) \in \mathbb{R}^2$.

In this special case it is actually also possible to eliminate s and t, and write z as a function of x and y. From the first two equations in (12.11) we see that

$$s = \ln(-x),$$
$$t = -\frac{y}{2x}.$$

If this is inserted into the third equation we obtain

$$z = \ln(-x) + \left(\frac{y}{2x}\right)^2.$$

This could also be obtained using the general solution found in Section 12.1, given by

$$z = \ln(-x) + G\left(\frac{y}{-x}\right), \qquad x < 0.$$

Applying the initial condition $x = -1$, $y = 2t$, and $z = t^2$, we get

$$t^2 = \ln(1) + G\left(\frac{2t}{1}\right) = G(2t).$$

This implies that the function G is defined as

$$G(t) = \left(\frac{1}{2}t\right)^2.$$

Hence, we obtain the solution

$$z = \ln(-x) + \left(\frac{y}{2x}\right)^2.$$

Recipe

1. Be sure that the PDE is written in the form

    $$P(x, y, z) \frac{\partial z}{\partial x} + Q(x, y, z) \frac{\partial z}{\partial y} - R(x, y, z) = 0,$$

    with no derivatives in P, Q, and R. Do not forget the minus sign in front of R!

2. Determine the Γ-curve and put it in the parametric form

    $$\mathbf{r}(s) = [x(s), y(s), z(s)], s \in \mathbb{R}.$$

    Often, it is possible to use x or y as the parameter, say, $x = x$, $y = g(x)$, $z = f(x)$.

3. Form the ODE system for the characteristic curves,

    $$\frac{\mathrm{d}x}{\mathrm{d}s} = P(x, y, z),$$

    $$\frac{\mathrm{d}y}{\mathrm{d}s} = Q(x, y, z),$$

    $$\frac{\mathrm{d}z}{\mathrm{d}s} = R(x, y, z),$$

    and solve it by some standard method (in real life, this may have to be carried out numerically).

4. Determine the collection of characteristics passing through Γ by imposing appropriate initial conditions on the solution found in the previous point. Make sure this defines a surface, which, as in the example, may sometimes be reduced to the explicit form $z = f(x, y)$.

12.3   System of Linear PDEs

We consider the system of the form

$$\sum_{j=1}^{n} a_{i,j}(x, y) u_x^{(j)} + b_{i,j}(x, y) u_y^{(j)} = c_i, \quad i = 1, \ldots, n, \tag{12.12}$$

where the unknown $u^{(1)}$, $u^{(2)}$,..., $u^{(n)}$ depend on two independent variables x and y. In the matrix form, the above system can be written as

$$A U_x + B U_y = c, \tag{12.13}$$

where $A = (a_{i,j}(x,y))$ and $B = (b_{i,j}(x,y))$ are matrices and $c = (c_i)$ and $U = (u^{(i)})$ are vectors. If the matrix A is invertible, then (12.13) can be rewritten as

$$U_x + DU_y = d, \qquad (12.14)$$

where $D = A^{-1}B$ and $d = A^{-1}c$. Let $\lambda_1, \ldots, \lambda_n$ be the eigenvalues of the matrix D. Assume that the system (12.14) is hyperbolic, that is, the matrix D has n distinct real eigenvalues. Then D is diagonalisable and can be decomposed as $D = P\Lambda P^{-1}$, where $P = (e_1, \ldots, e_n)$ is a matrix formed with the eigenvectors e_i associated to λ_i, and $\Lambda = Diag(\lambda_1, \ldots, \lambda_n)$ is a diagonal matrix formed with the eigenvalues. Setting $V = P^{-1}U$ (i.e., $U = PV$), the system (12.14) becomes

$$V_x + \Lambda V_y = q, \qquad (12.15)$$

where $q = P^{-1}d - P^{-1}P_xV - P^{-1}DP_yV$. Component-wise, the system (12.15) can be broken down as follows:

$$V_x^{(i)} + \lambda_i V_y^{(i)} = q_i, \quad i = 1, \ldots, n.$$

The system (12.15) is called the canonical form of the system (12.14).

For example, consider the system

$$u_x^{(1)} + (2x - 2y)u_y^{(2)} + (4y - x)u_y^{(1)} - 3xy = 0,$$
$$u_x^{(2)} + (2y - 2x)u_y^{(1)} + (4x - y)u_y^{(2)} - 6xy = 0.$$

In the matrix form, it can be written as

$$U_x + AU_y = c,$$

where $A = \begin{pmatrix} 4y - x & 2x - 2y \\ 2y - 2x & 4x - y \end{pmatrix}$, $U = \begin{pmatrix} u^{(1)} \\ u^{(2)} \end{pmatrix}$ and $c = \begin{pmatrix} 3xy \\ 6xy \end{pmatrix}$. The matrix A has two eigenvalues $\lambda_1 = 3x$ and $\lambda_2 = 3y$. The system is hyperbolic in the region $\Omega = \{(x,y) \in \mathbb{R}^2 | x \neq y\}$. In this region, $A = P\Lambda P^{-1}$, where $P = \begin{pmatrix} 1 & 2 \\ 2 & 1 \end{pmatrix}$ and $\Lambda = \begin{pmatrix} 3x & 0 \\ 0 & 3y \end{pmatrix}$. The canonical form of the system is

$$V_x + \Lambda V_y = q,$$

where $V = P^{-1}U$ and $q = P^{-1}c = \begin{pmatrix} 3xy \\ 0 \end{pmatrix}$. In terms of components, we have

$$v_x^{(1)} + 3xv_y^{(1)} = 3xy,$$
$$v_x^{(2)} + 3yv_y^{(2)} = 0,$$

with general solution

$$v^{(1)} = \frac{1}{2}y^2 + f_1\left(y - \frac{3}{2}x^2\right),$$
$$v^{(2)} = f_2(ye^{-3x}),$$

where f_1 and f_2 are arbitrary functions. Since $U = PV$, we obtain

$$
u^{(1)} \;=\; \frac{1}{2}y^2 + f_1\left(y - \frac{3}{2}x^2\right) + 2f_2(ye^{-3x}),
$$

$$
u^{(2)} \;=\; y^2 + 2f_1\left(y - \frac{3}{2}x^2\right) + f_2(ye^{-3x}).
$$

If we consider the initial condition $U = \begin{pmatrix} 2y^2 \\ 3y^2 \end{pmatrix}$ on $x = 0$, then $f_1(y) = \frac{5}{6}y^2$ and $f_2(y) = \frac{1}{3}y^2$. The particular solution is given by

$$
u^{(1)} \;=\; \frac{1}{2}y^2 + \frac{5}{6}\left(y - \frac{3}{2}x^2\right)^2 + \frac{2}{3}(ye^{-3x})^2,
$$

$$
u^{(2)} \;=\; y^2 + \frac{5}{3}\left(y - \frac{3}{2}x^2\right)^2 + \frac{1}{3}(ye^{-3x})^2.
$$

12.4 System of Quasilinear PDEs

We now consider the system of the form

$$
U_x + D(x, y, U)U_y = d(x, y, U), \tag{12.16}
$$

where D is a matrix, $U = U(x, y) = (u^{(1)}, \ldots, u^{(n)})$ and $d = (d_1, \ldots, d_n)$ are vectors. We restrict the analysis to hyperbolic systems. Then $D = P\Lambda P^{-1}$, where $\Lambda = \Lambda(x, y, U)$ is a diagonal matrix formed of the eigenvalues of D and P is a matrix formed of the eigenvectors of D. Then the system (12.16) can be written in the normal form

$$
P^{-1}U_x + \Lambda P^{-1}U_y = P^{-1}d. \tag{12.17}
$$

Component-wise, it is written as

$$
\sum_{j=1}^{n} P_{i,j}^{-1}\left(\frac{\partial}{\partial x}u^{(j)} + \lambda_i \frac{\partial}{\partial y}u^{(j)}\right) = \sum_{j=1}^{n} P_{i,j}^{-1}d_j, \quad i = 1, \ldots, n, \tag{12.18}
$$

where the λ_i are eigenvalues on the i-th row of the matrix Λ. The i-th characteristic is the curve in the (x, y)-plane satisfying

$$
\frac{dx}{ds_i} = 1,
$$

$$
\frac{dy}{ds_i} = \lambda_i,
$$

or equivalently,

$$\frac{dx}{dy} = \lambda_i.$$

Since the directional derivative parallel to the i-th characteristic is

$$\frac{d}{ds_i}u^{(j)} = \frac{\partial}{\partial x}u^{(j)}\frac{dx}{ds_i} + \frac{\partial}{\partial y}u^{(j)}\frac{dy}{dx} = \frac{\partial}{\partial x}u^{(j)} + \lambda_i\frac{\partial}{\partial y}u^{(j)},$$

then the PDE system (12.18) in the normal form reduces to the system of ODEs

$$\sum_{j=1}^{n} P_{i,j}^{-1}\frac{d}{ds_i}u^{(j)} = \sum_{j=1}^{n} P_{i,j}^{-1}d_j, \quad i = 1, \ldots, n.$$

As an application, consider the one-dimensional isentropic flow of an inviscid gas governed by the system

$$\frac{\partial u}{\partial t} + u\frac{\partial u}{\partial x} + \frac{c^2}{\rho}\frac{\partial \rho}{\partial x} = 0, \quad \text{(Euler's equation)}$$

$$\frac{\partial \rho}{\partial t} + \rho\frac{\partial u}{\partial x} + u\frac{\partial \rho}{\partial x} = 0, \quad \text{(continuity equation)}$$

where $u(x,t)$ and $\rho(x,t)$ are the velocity and density of the gas, respectively, c is the speed of sound. Further information on the derivation of these equations can be found in [16]. In matrix form, the system can be written as

$$\frac{\partial w}{\partial t} + D\frac{\partial w}{\partial x} = 0,$$

where $w = \begin{pmatrix} u \\ \rho \end{pmatrix}$ and $D = \begin{pmatrix} u & \frac{c^2}{\rho} \\ \rho & u \end{pmatrix}$. The matrix D has two eigenvalues $\lambda_1 = u + c$ and $\lambda_2 = u - c$. The system is hence hyperbolic and $D = P\Lambda P^{-1}$, with

$$P = \begin{pmatrix} c & c \\ \rho & -\rho \end{pmatrix}, \quad P^{-1} = -\frac{1}{2c\rho}\begin{pmatrix} -\rho & -c \\ -\rho & c \end{pmatrix}, \quad \Lambda = \begin{pmatrix} u+c & 0 \\ 0 & u-c \end{pmatrix}.$$

Along characteristic curves, we have

$$\frac{dt}{ds_1} = 1, \quad \frac{dx}{ds_1} = u + c,$$

$$\frac{dt}{ds_2} = 1, \quad \frac{dx}{ds_2} = u - c.$$

Then, the reduced canonical form is given by the system

$$\rho\frac{du}{ds_1} + c\frac{d\rho}{ds_1} = 0,$$

$$\rho\frac{du}{ds_2} - c\frac{d\rho}{ds_2} = 0.$$

Its general solution is

$$\frac{u}{c} + \ln \rho = f\left(\ln \rho - \frac{u}{c}\right),$$

where f is an arbitrary function of one argument.

12.5   Exercises

1. Find the solution surface, $z = f(x, y)$, to the quasi-linear first order partial differential equation

$$\frac{\partial z}{\partial x} + y\frac{\partial z}{\partial y} - 2 = 0$$

   so that the space curve defined by

$$x = t,$$
$$y = 1,$$
$$z = t,$$

   $t \in \mathbb{R}$ is in the surface. *Hint:* Follow the recipe step-by-step.

2. Given the partial differential equation

$$3yu_x - 2yu_y = 0,$$

   where $u_x = \partial u(x, y)/\partial x$ etc. Solve the characteristic equations and sketch the graph of the characteristics. State the general solution of the equation and try to find solutions satisfying the following Cauchy Data on the given curves:

   (a) $u(x, y) = x^2$ on the line $y = x$.
   (b) $u(x, y) = x^2$ on the line $y = -x$.
   (c) $u(x, y) = 2x$ on the ellipse $3y^2 + 2x = 4$.

3. Consider the following quasi-linear partial differential equations, and use the method of characteristics to find solutions passing through the given curves τ (the solutions may only be defined implicitly):

   (a) $yu_x - xu_y = e^u$; τ defined by $y = \sin x$, $u = 0$.
   (b) $xu_x + yu_y = \sec u$; τ defined by $x = s^2$, $y = \sin s$, $u = 0$, $s \in \mathbb{R}$.
   (c) $u_x - xu_y = 4$; τ defined by $y = 4x$, $u = 0$.

4. Consider the Cauchy problem

$$\frac{\partial \rho}{\partial t} + \frac{\partial}{\partial x}(g(x)\rho), t > 0, x \in \mathbb{R},$$
$$\rho(0, x) = \rho^0(x),$$

where the function g is given. The equation can describe the evolution of the density $\rho(t, x)$ of metastatic tumors of size x at time t, and the function g be seen as the tumor expansion/growth rate.

(a) Determine the characteristic curves when $g(x) = x$ (exponential growth) and $g(x) = ax \ln \left(\frac{b}{x}\right)$ (Gompertzian growth).

(b) Deduce the solution of the Cauchy problem when $g(x) = x$.

(c) Determine the solution of the Cauchy problem when $g(x) = ax \ln \left(\frac{b}{x}\right)$.
    Hint: Show that the function $W(t) = g(x(t))\rho(t, x)$ is constant along the characteristic curves.

13

Solution Hints

Exercise 2.5.1: Acceleration, $[a] = \text{ms}^{-2}$; mass density, $[\rho] = \text{kgm}^{-3}$; electric power, $[W] = \text{kgm}^2\text{s}^{-3}$; air pressure $[p] = \text{kgm}^{-1}\text{s}^{-2}$; specific heat capacity $[c] = \text{kg}^2\text{m}^2\text{s}^{-3}\text{K}^{-1}$; heat conduction coefficient, $[k] = \text{kgms}^{-3}\text{K}^{-1}$.

Exercise 2.5.3: The rank of the matrix is 3, and we need 3 core variables. There are $\binom{6}{3} = 20$ possible subsets of 3 variables of which $\{R_1, R_2, R_3\}$ and $\{R_1, R_2, R_4\}$ are two of a total of 4(?) possible sets of core variables.

Exercise 2.5.5: (a) We have 5 physical quantities, F_d, A, U, ρ_{air}, v, and the 3×5 dimensional matrix has rank 3, hence there are 2 dimensionless variables. Applying A, U, ρ as core variables, it follows that we may write $\pi_1 = \frac{U\sqrt{A}}{v}$ and $\pi_2 = \frac{F_d}{\rho A U^2}$, and the relation in the text follows.

(b) At constant speed, $F_d = mg$. Setting $\phi \equiv 1$, we may solve for U, $U = \sqrt{2mg/(\rho_{air} A)}$.

(c) Taking, say, $m = 75\text{kg}$, $A = 0.5\text{m}^2$, $\rho_{air} = 1.2\text{kg/m}^3$, and $g = 9.8\text{m/s}^2$, we find $U \simeq 50\text{m/s} \approx 180\text{km/h}$.

Exercise 2.5.7: The straightforward 3×6 dimension matrix for the 6 quantities F, U, d, ρ, ω, and μ has rank 3 and we have 3 dimensionless variables. With U, d, and ρ as core variables, we obtain $\pi_1 = \frac{\omega d}{U}$, $\pi_2 = \frac{U d \rho}{\mu}$, and $\pi_3 = \frac{F}{\rho U^2 d^2}$ and the given relation follows.

Exercise 2.5.9: The relation $\omega = f(l, \rho, F)$ involves 4 quantities, and the 3×4 dimensional matrix has rank 1. Since total mass M of the band remains constant, $M = l_0 \rho_0 = l\rho$, and $\rho = \frac{l_0}{l}\rho_0$. Selecting l, F and ρ as core variables, we obtain the dimensionless variable $\pi = \frac{1}{\omega}\sqrt{\frac{F}{\rho l^2}}$, from which we have

$$\omega = C\sqrt{\frac{F}{\rho l^2}},$$

DOI: 10.1201/9781003725206-13

where C is an unknown constant. Now introducing $F = F_0 \frac{l-l_0}{l_0}$ and $\rho = \frac{l_0}{l}\rho_0$,

$$\omega = C\sqrt{\frac{F_0 \frac{l-l_0}{l_0}}{\frac{l_0}{l}\rho_0 l^2}} = C\sqrt{\frac{F_0}{\rho_0 l_0^2}}\sqrt{1 - \frac{l_0}{l}}.$$

When $l \gg l_0$, the frequency is almost constant and equal to $C\sqrt{\frac{F_0}{\rho_0 l_0^2}}$ (Obviously, C depends on the material constants for rubber, *e.g.*, for elasticity).

Exercise 3.5.1: A reasonable time scale may be obtained from

$$T = \frac{\max|U^*|}{\max\left|\frac{dU^*}{dt^*}\right|} = \frac{A}{A2\pi f_0} = \frac{1}{2\pi f_0},$$

which is the second item on the list. Both the first and the third combinations are acceptable, whereas the last one is clearly too long to be reasonable.

Exercise 3.5.3: Applying Lin and Segel's rule of thumb,

$$T_1 = \frac{\max|u|}{\max\left|\frac{du}{dt}\right|} = \frac{A}{Aa} = \frac{1}{a}.$$

The tangent slope at $(t_0, u(t_0))$ is $\frac{du}{dt}(t_0) = -aAe^{-at_0}$. The crossing point on the t-axis is obtained from the equation

$$Ae^{at_0} - aAe^{-at_0}(t_1 - t_0) = 0,$$

that is, $T_2 = t_1 - t_0 = 1/a$, which gives the same answer as the rule.

Exercise 4.1.3.1: (a) We apply the regular perturbation by introducing $x(t,\varepsilon) = x_0(t) + \varepsilon x_1(t) + \varepsilon^2 x_2(t) + \cdots$, leading to the recursive system

$$\begin{aligned}
2x_0'' &= 1, & x_0(0) &= 0, \ x_0'(0) = 0, \\
2x_1'' + x_0' &= 0, & x_1(0) &= 0, \ x_1'(0) = 0, \\
2x_2'' + x_1'' &= 0, & x_2(0) &= 0, \ x_2'(0) = 0, \\
& \cdots & & \cdots
\end{aligned}$$

The solution is straightforward, $x_0(t) = t^2/4$, $x_1(t) = -t^3/24$, $x_2(t) = t^4/192$, and we obtain an infinite series for the solution

$$x_{\text{sol}}(t) = \frac{t^2}{4} - \varepsilon\frac{t^3}{24} + \varepsilon^2\frac{t^4}{192} + \cdots.$$

The start of the expansion in ε for the exact solution follows after applying $e^x = 1 + x + \frac{x^2}{2!} + \frac{x^3}{3!} + \cdots$ for $e^{-\varepsilon\frac{t}{2}}$. The series is convergent for all fixed values of ε and t **(b)** When $t \in [0,1]$ and $x_a(t,\varepsilon) = \frac{t^2}{4} - \varepsilon\frac{t^3}{24}$, we may write

$$|x_a(t,\varepsilon) - x_{\text{sol}}(t,\varepsilon)| = \left|\frac{t^2}{4} - \varepsilon\frac{t^3}{24} - x_{\text{sol}}(t,\varepsilon)\right| < M\varepsilon^2$$

where M is a constant which together with ε is taking care of the remaining terms. Letting $\varepsilon \to 0$, this is clearly a uniform approximation. This is not possible if $[0,1]$ is replaced with $[0,\infty]$: For a fixed ε, $x_a(t,\varepsilon) \sim -\varepsilon \frac{t^3}{24}$ for large $|t|$-s. However, $x_{sol}(t,\varepsilon) \sim \frac{t}{\varepsilon}$, for large $|t|$-s, and $\max_{t\in[0,\infty]} |x_a(t,\varepsilon) - x_{sol}(t,\varepsilon)| = \infty$ for all $\varepsilon > 0$.

Exercise 4.1.3.3: (a) The scaling is $v^* = V_0 v$, and $t^* = \frac{m}{a}t$, thus,

$$\frac{dv}{dt} = -v + \varepsilon v^2, v(0) = 1,$$

$$\varepsilon = bV_0/a \ll 1.$$

(b) Inserting $v(t) = v_0(t) + \varepsilon v_1(t) + \cdots$ leads to the system

$$v_0' + v_0 = 0, \qquad v_0(0) = 1,$$
$$v_1' + v_1 - v_0^2 = 0, \quad v_1(0) = 0,$$

and the solutions $v_0(t) = e^{-t}$ and $v_1(t) = e^{-t} - e^{-2t}$. Thus,

$$v(t) = e^{-t} + \varepsilon\left(e^{-t} - e^{-2t}\right) + O\left(\varepsilon^2\right).$$

The exact solution, where $v_{ex}(0) = 1$, is

$$v_{ex}(t) = \frac{e^{-t}}{1 + \varepsilon(e^{-t} - 1)}.$$

Since $\varepsilon |e^{-t} - 1| \ll 1$ for $t \geq 0$, we have

$$v(t) = \frac{e^{-t}}{1 + \varepsilon(e^{-t} - 1)} = e^{-t}\left[1 - \varepsilon\left(e^{-t} - 1\right) + O\left(\varepsilon^2\right)\right]$$
$$= e^{-t} - \varepsilon\left(e^{-2t} - e^{-t}\right) + O\left(\varepsilon^2\right).$$

Exercise 4.2.6.1: (a) Since $0 < \varepsilon \ll 1$, there are clearly 2 roots around $x = 1$, and there must be one big positive and one big negative root remaining because εx^4 dominates for $|x| \gg 1$. Further approximations to the two roots around $x = 1$ may be found by regular perturbation. For the other pair, introduce $x = a/\varepsilon^\delta$, where we find by inspection that $\delta = 1/2$ leads to the balanced equation

$$a^4 - a^2 + a^3\sqrt{\varepsilon} - \varepsilon + 2a\sqrt{\varepsilon} = 0,$$

with the leading order equation $a^4 - a^2 = 0$. Hence, to the leading order, the 4 solutions are $-\varepsilon^{-1/2}, 1, 1$, and $\varepsilon^{-1/2}$.

(b) In this case, there is one root around 2. It is the only real one since the slope of $y = \varepsilon x^3 + x - 2$ is always larger than 1. However, there are complex roots, and assuming, as in (a), $x = a/\varepsilon^\delta$, it turns out that the balanced equation for $\delta = \sqrt{\varepsilon}$ becomes $a^3 + a - 2\sqrt{\varepsilon} = 0$ with leading order solutions $a = 0$, corresponding to the solution 2, and $a = \pm i$. Leading order solutions then become $x = 2, \pm i\varepsilon^{-1/2}$.

(c) Carry out this investigation by yourself.

Exercise 4.2.6.3: The general solution is $u(x) = A\cos\omega t + B\sin\omega t$, $\omega = \varepsilon^{-1/2}$. The outer solution, $u(x) \equiv 0$, is nowhere near the full solution. Moreover, the problem does not even have solutions when $2\pi n = \varepsilon^{-1/2}$, since then $u(1) = u(2)$.

Exercise 4.2.6.5: This is a tricky question. The outer solution, $y_{\text{out}}(x) = \exp(-x^2/2)$ happens to satisfy both initial conditions, and there is no boundary layer near $x = 0$ for these special initial conditions. In fact, it seems that the full solution converges uniformly to the outer solution when $\varepsilon \to 0$, which is typical for a regular perturbation problem. For large x, the second term becomes small, and we expect an oscillatory behaviour, $\sim A(x, \varepsilon) \cos(\varepsilon^{-1/2}x + \phi)$, where $A(x)$ is a slowly varying function in x. Actually, for $\varepsilon = 1$, the analytical solution is the Bessel function $J_0(x)$, and the exact solution is similar, with an amplitude for the wiggle, $|A|$, tending to 0 when $\varepsilon \to 0$.

Exercise 4.2.6.7: (a) With time scale $T = 1/\omega$, and A as scale for c^*, it follows by inspection of (3.66) and (3.67), that the scale for n^* should be

$$N_0 = \frac{\delta}{\beta}A,$$

and the system in (3.67) follows. The remaining parameters are

$$\kappa = \frac{\alpha_0}{\omega} = \frac{1/\omega}{1/\alpha_0},$$

$$\varepsilon = \frac{\omega}{\delta} = \frac{1/\delta}{1/\omega}.$$

Both are ratios of time scales where it is assumed that κ somewhat larger than 1, and $0 < \varepsilon \ll 1$. This is therefore a singularly perturbed system. The equilibrium points may easily be read from the equation:

$$(n_1, c_1) = (0, 0),$$
$$(n_2, c_2) = (\kappa - 1, \kappa - 1).$$

Linearization around $(0, 0)$ gives

$$\begin{bmatrix} \kappa - 1 & 0 \\ \frac{1}{\varepsilon} & -\frac{1}{\varepsilon} \end{bmatrix}.$$

The eigenvalues are $\lambda_1 = \kappa - 1 > 0$ and $\lambda_2 = -1/\varepsilon < 0$. Thus, the point becomes a saddle point.

(b) Eq. (3.67) is a singular perturbation system since $\varepsilon \ll 1$. The outer solution to leading order implies $n_0(t) = c_0(t)$, and this gives

$$\frac{dn_0}{dt} = \left(\frac{\kappa}{1+n_0} - 1\right) n_0. \tag{13.1}$$

The point $n_2 = c_2 = \kappa - 1$ is still an equilibrium point, and it follows from an inspection of (13.1) that it is stable regardless of where we choose to start. The path for the leading order outer solution is now simply the line $n_0 = c_0$. Observe that the implicit solution of (13.1) is

$$\frac{n_0(t)}{|n_0(t) - (\kappa - 1)|^\kappa} = \exp\left[(t - t_0)(\kappa - 1)\right].$$

(c) With the new time scale ε, and $\tau = t/\varepsilon$, we obtain the following system for the leading order inner solution $N_0(\tau)$ and $C_0(\tau)$

$$\frac{dN_0}{d\tau} = 0,$$

$$\frac{dC_0}{d\tau} = N_0 - C_0.$$

Starting at $\{n(0), c(0)\}$ leads to

$$N_0(\tau) = n(0),$$
$$C_0(\tau) = [c(0) - n(0)]\, e^{-\tau} + n(0).$$

By applying the matching conditions, we obtain the uniform solution

$$n^u(t) = n_0(t),$$
$$c^u(t) = n_0(t) + [c(0) - n(0)]\, e^{-t/\varepsilon}.$$

Exercise 5.5.1: With $f(u) = u(25 - u)$, $u_0 = 0$, and $u_1 = 25$ are equilibrium solutions. Moreover, u_0 is unstable since $f'(0) = 25$, and u_1 is stable since $f'(25) = -2$.

Exercise 5.5.3: The sign analysis is described in the text. The important observation is that $f(\mu, u)$ can only change sign when u moves through an equilibrium, thus defining regions in the (μ, u)-plane where $f(\mu, u)$ has the same sign.
i) The equilibrium points occur for $(\mu + u)(\mu + u^2 - 2u - 1) = 0$, that is, for $u_0 = -\mu$, or $u_0^2 - 2u_0 - 1 = -\mu$. In the (μ, u)-plane the first represents a

line and the second a parabola with vertex $(1, 2)$ that opens to the left. Make a drawing, identify the regions with constant sign, and finally determine the stable and unstable equilibrium points by moving u vertically past the equilibrium point.

ii) In this case the collection of equilibrium points is on the curves $u = 0, \mu = 9/u$, or $\mu = u^2 - 2u$. The analysis is similar to i).

Exercise 5.5.5 (a): Since C is a constant, setting $Q_0 - \sigma T^4 = 0$, the equilibrium point is $T_0 = \sqrt[4]{\frac{Q_0}{\sigma}}$, and since $\frac{d}{dt}(Q_0 - \sigma T^4) = -4\sigma T_0^3 < 0$, the point is stable.

The time scale is found by linearizing the equation around the equilibrium solution. Set $T = T_0 + y$. Then the linearized equation becomes

$$\dot{y} = -\frac{4\sigma T_0^3}{C} y,$$

and $\frac{C}{4\sigma T_0^3}$ becomes a reasonable time scale. With the numbers in the text,

$$\frac{C}{4\sigma T_0^3} = \frac{1}{4} \times 6 \times 10^9 \text{daysK}^3 \frac{1}{(287\text{K})^3} \approx 63\text{days}.$$

(b) Consider the RHS of the extended model, where the stationary points are solutions of

$$Q_a \tanh\left(\frac{T - T_0}{T_n}\right) = \sigma T^4 - Q_0.$$

One solution is T_0, whereas two others exist when the slope of the left-hand side at $T = T_0$ is larger than the slope of the right-hand side,

$$\frac{Q_a}{T_n} > 4\sigma T_0^3,$$

This corresponds to the inequality stated in (b). What can be said about the stability of the equilibrium points?

Exercise 5.5.7: There are many possible solutions, but one of the simplest is probably

$$\dot{x} = (x + 1)\, x,$$
$$\dot{y} = (y - 3)(y - 5).$$

Exercise 5.5.9: We note that $(0, 0)$ is a stationary point and determine the eigenvalues of the coefficient matrix A from the equation $\det(A - \lambda I) = (1 - \lambda)^2 - \mu^2 = 0$. Hence, $\lambda_{1,2} = 1 \pm \mu$, and $(0, 0)$ is always unstable. This is an unstable node for $|\mu| < 1$, changing to a saddle for $|\mu| > 1$.

Exercise 6.4.1: (a) It is a logistic population model with a harvesting term, $-\alpha N^*$. We may take N_0 as a scale for N^* as long as $N_0 \gg K$, and the scale r^{-1} for time as long as $\alpha < r$. These scales only apply to the initial behaviour of the system.
(b) The equations model two interacting populations, each of which follows a logistic model in the absence of the other. Each of the two populations is both predator and prey. The parameters α and β measure the intensity of the predator/prey interaction. Let $\beta = 0$. It is easily seen that 0 and 1 are equilibria for y (now uncoupled from x). If $y = 0$, equilibria points for x are the well-known 0 and 1. Thus, both $(0,0)$ and $(1,0)$ will be unstable. Finally, if $y = 1$, 0 and $1 - \alpha$ are stationary points for x. Hence, $(0,1)$ is unstable because of x, whereas $(1 - \alpha, 1)$ is stable.

Exercise 6.4.3: (a) The model is basically a logistic population model with stationary points $P_1 = 0$, $P_2 = m$, and $P_3 = M$. However, if the population drops below m, it dies out. The stability of the equilibrium points follows easily from a sketch of the RHS of the equation.
(b) The equilibrium points (P_0, J_0) of the simplified model are $(0,0)$, $(1,0)$, and $\left(\frac{1}{2}, \frac{1}{4}\right)$. After forming the Jacobian, it is easily seen that $(0,0)$ are $(1,0)$ unstable saddle points, whereas the Jacobian for $\left(\frac{1}{2}, \frac{1}{4}\right)$ is

$$\begin{bmatrix} 0 & -1 \\ \frac{1}{4} & 0 \end{bmatrix},$$

with eigenvalues, $\pm i/2$. Thus, the point appears to be a centre.
(c) The hint is somewhat tricky to apply:

$$\nabla h\left(P, J\right) \cdot \left(\frac{dP}{dt}, \frac{dJ}{dt}\right) = -\frac{3}{2}\left(1 - 2P\right)\left(P\left(1 - P\right) - J\right) + 1 \cdot \left(-\frac{J}{2} + JP\right).$$

At this stage, we solve for J in $h\left(P, J\right) = 0$, that is

$$J = \frac{3}{2}P\left(1 - P\right),$$

and then insert into the expression and obtain 0. **(d)** After introducing $x = P - 1/2$ and $y = J - 1/4$, we obtain the system of differential equations

$$\dot{x} = -x^2 - y$$
$$\dot{y} = x\left(y + \frac{1}{4}\right)$$

(e) One can observe the following properties immediately from the model

- The presence of hunters reduces the number of animals.
- If $P > \frac{1}{2}$, J will increase, and if $P < \frac{1}{2}$, J decreases.

Exercise 13.2.1: The characteristic system is

$$\frac{dx}{ds} = 1,$$

$$\frac{dy}{ds} = y,$$

$$\frac{dz}{ds} = 2,$$

with general solution $x(s) = s + A$, $y(s) = Be^s$, $z(s) = 2s + C$ for arbitrary constants A, B, and C. The space curve Γ is $\mathbf{r}(s) = t\mathbf{x} + \mathbf{j} + t\mathbf{k}$, and the characteristics cross Γ at $s = 0$. Hence, the solution surface becomes (in parametric form)

$$x(s, t) = t + s,$$

$$y(s, t) = e^s,$$

$$z(s, t) = t + 2s.$$

Eliminating the parameters t and s, we may write the solution

$$z = \ln y + x.$$

Exercise 13.2.3: (a) The characteristic curves are obtained from

$$\frac{dx}{dt} = y, \frac{dy}{dt} = -x, \frac{du}{dt} = e^u,$$

resulting in

$$x(t) = a \cos t + b \sin t,$$

$$y(t) = b \cos t - a \sin t,$$

$$u(t) = -\ln(t_0 - t).$$

In order to find a, b, t_0, we parametrize the curve τ as $x = s$, $y = \sin s$, $u = 0$. The curves have a common point for $t = 0$ if $t_0 = 1$. Moreover, it follows that $a = s$, $b = \sin s$. The solution, parametrized in terms of (s, t), thus becomes

$$x(t, s) = s \cos t + \sin s \sin t,$$

$$y(t, s) = \sin s \cos t - s \sin t,$$

$$u(t, s) = -\ln(1 - t).$$

By applying the identities

$$s = x \cos t - y \sin t,$$

$$\sin s = y \cos t + x \sin t,$$

it is easy to eliminate t and s and obtain the implicit solution

$$\sin\left[x\cos\left(1-e^{-u}\right)-y\sin\left(1-e^{-u}\right)\right]=y\cos\left(1-e^{-u}\right)+x\sin\left(1-e^{-u}\right).$$

(b) The characteristic equations are

$$\frac{dx}{dt}=x,\ \frac{dy}{dt}=y,\ \frac{du}{dt}=\frac{1}{\cos u}$$

with (partly implicit) solutions

$$x(t)=ae^{t},$$
$$y(t)=be^{t},$$
$$\sin u=t+c.$$

Similar to above, τ is parametrized as $x=s^{2}$, $y=\sin s$, $u=0$, and the solution becomes

$$x(t)=s^{2}e^{t},$$
$$y(t)=\sin se^{t},$$
$$\sin u=t.$$

Here, x may be expressed in terms of y and u as

$$x=\left(\sin^{-1}\left(ye^{-\sin u}\right)\right)^{2}e^{\sin u}.$$

(c) The solution follows similarly to (a) and (b) above.

Bibliography

[1] Warder C Allee and Edith S Bowen. Studies in animal aggregations: mass protection against colloidal silver among goldfishes. *Journal of Experimental Zoology*, 61(2):185–207, 1932.

[2] Warder Clyde Allee. The social life of animals. 1938.

[3] Warder Clyde Allee, Orlando Park, Alfred E Emerson, Thomas Park, and Karl P Schmidt. *Principles of animal ecology*. Number Edn 1. 1949.

[4] Linda JS Allen, Fred Brauer, Pauline Van den Driessche, and Jianhong Wu. *Mathematical epidemiology*, volume 1945. Springer, 2008.

[5] RM Anderson and RM May. *Infectious Diseases of Humans*. Oxford University Press, 1991.

[6] Grigory Isaakovich Barenblatt. *Scaling, self-similarity, and intermediate asymptotics: dimensional analysis and intermediate asymptotics*. Number 14. Cambridge University Press, 1996.

[7] Carl M Bender and Steven A Orszag. *Advanced mathematical methods for scientists and engineers I: Asymptotic methods and perturbation theory*. Springer Science & Business Media, 2013.

[8] D Bernoulli. Essai d'une nouvelle analyse de la mortalité causée par la petite vérole et des avantages de l'inoculation pour la prévenir. *Hist. Acad. R. Sci. Paris*, pages 1–45, 1760.

[9] Tomas Bohr, Peter Dimon, and Vakhtang Putkaradze. Shallow-water approach to the circular hydraulic jump. *Journal of Fluid Mechanics*, 254:635–648, 1993.

[10] PW Bridgman. Dimensional analysis, yale, 1922.

[11] Earl A Coddington and Norman Levinson. *Theory of ordinary differential equations*. McGraw-Hill New York, 1955.

[12] John Crank. *The mathematics of diffusion*. Oxford university press, 1979.

[13] Mohammed Y Dawed, Patrick M Tchepmo Djomegni, and Harald E Krogstad. Complex dynamics in a tritrophic food chain model with general functional response. *Natural Resource Modeling*, 33(2):e12260, 2020.

[14] PM Tchepmo Djomegni and KS Govinder. *The interplay of dynamical systems analysis and group theory.* Thesis (M.Sc.)-University of KwaZulu-Natal, 2011.

[15] Neville D Fowkes, John J Mahony, and Neville de Mestre. An introduction to mathematical modelling.

[16] Paul R Garabedian. *Partial differential equations,* volume 325. American Mathematical Society, 2023.

[17] NH Gartner, Carroll J Messer, and Ajay K Rathi. Traffic flow theory (update of trb special report 1165). *TRB, Washington, DC, USA, Tech. Rep,* 1165, 1997.

[18] GF Gause. The struggle for existence williams and wilkins. *Baltimore, Maryland,* 1934.

[19] H Scott Gordon. The economic theory of a common-property resource: the fishery. *Journal of Political Economy,* 62(2):124–142, 1954.

[20] P Grindrod. Patterns and waves [m], 1991.

[21] David M Grobman. Homeomorphism of systems of differential equations. *Doklady Akademii Nauk SSSR,* 128(5):880–881, 1959.

[22] Richard Haberman. *Mathematical models: mechanical vibrations, population dynamics, and traffic flow.* SIAM, 1998.

[23] Philip Hartman. On local homeomorphisms of euclidean spaces. *Bol. Soc. Mat. Mexicana,* 5(2):220–241, 1960.

[24] M Hirsch and S Smale. *Differential Equations, Dynamical Systems, and Linear Algebra.* Academic Press, 1974.

[25] Dominic Jordan and Peter Smith. *Nonlinear ordinary differential equations: an introduction for scientists and engineers.* OUP Oxford, 2007.

[26] WO Kermack and AG McKendrick. A contribution to the mathematical theory of epidemics. *Proceedings of the Royal Society of London, Series,* 115:700–721, 1927.

[27] Harald E Krogstad, Mohammed Yiha Dawed, and Tadele Tesfa Tegegne. Alternative analysis of the michaelis–menten equations. *Teaching Mathematics and its Applications: An International Journal of the IMA,* 30(3):138–146, 2011.

[28] H Lamb. Hydrodynamics. Cambridge University Press, 1932. *Burnham, DC and Hallock, JN: Chicago Monostatic Acoustic Vortex Sensing System,* 4:395–411, 1945.

[29] Michael James Lighthill and Gerald Beresford Whitham. On kinematic waves ii. a theory of traffic flow on long crowded roads. *Proceedings of the Royal Society of London. Series a. Mathematical and Physical Sciences*, 229(1178):317–345, 1955.

[30] Chia-Ch'iao Lin and Lee A Segel. *Mathematics applied to Deterministic Problems in the Natural Sciences*. SIAM, 1988.

[31] J Logan. *Applied Mathematics—A Contemporary Approach*. John Wiley and Sons, 1987.

[32] Thomas Robert Malthus. An essay on the principle of population (1798). *The Works of Thomas Robert Malthus, London, Pickering & Chatto Publishers*, 1:1–139, 1986.

[33] Robert M May, John R Beddington, Colin W Clark, Sidney J Holt, and Richard M Laws. Management of multispecies fisheries. *Science*, 205(4403):267–277, 1979.

[34] MB Schaefer. Some aspects of the dynamics of populations important to the management of the commercial fisheries. *InterAm Trop Tuna Comm Bull*, 1:25–56, 1954.

[35] BR Munson, DF Young, and TH Okiishi. *Fundamentals of Fluid Mechanics*. Wiley, 2002.

[36] PL Sachdev. Self-similarity and beyond: exact solutions of nonlinear problems. *Taylor & Francis eBooks DRM Free Collection*, 2000.

[37] Stephen G Simpson. Which set existence axioms are needed to prove the cauchy/peano theorem for ordinary differential equations? *The Journal of Symbolic Logic*, 49(3):783–802, 1984.

[38] Pierre-François Verhulst. Notice sur la loi que la population suit dans son accroissement. *Correspondence mathematique et physique*, 10:113–129, 1838.

[39] EJ Watson. The radial spread of a liquid jet over a horizontal plane. *Journal of Fluid Mechanics*, 20(3):481–499, 1964.

[40] Darcy Wentworth Thompson. On growth and form. 1917.

[41] GB Whitham. Linear and nonlinear waves. *Wiley-Interscience*, pages 431–484, 1974.

[42] Eleftherios C Zachmanoglou and Dale W Thoe. *Introduction to Partial Differential Equations with Applications*. Courier Corporation, 1986.

Index

For Product Safety Concerns and Information please contact our EU
representative GPSR@taylorandfrancis.com
Taylor & Francis Verlag GmbH, Kaufingerstraße 24, 80331 München, Germany

www.ingramcontent.com/pod-product-compliance
Lightning Source LLC
Chambersburg PA
CBHW060827170526
45158CB00001B/103